William MacCormac

Antiseptic Surgery

An Address Delivered at St. Thomas's Hospital....

William MacCormac

Antiseptic Surgery
An Address Delivered at St. Thomas's Hospital....

ISBN/EAN: 9783337021443

Printed in Europe, USA, Canada, Australia, Japan

Cover: Foto ©berggeist007 / pixelio.de

More available books at **www.hansebooks.com**

ANTISEPTIC SURGERY

AN ADDRESS

DELIVERED AT ST THOMAS'S HOSPITAL

WITH THE SUBSEQUENT DEBATE

TO WHICH ARE ADDED

A SHORT STATEMENT OF THE THEORY OF THE ANTISEPTIC METHOD, A DESCRIPTION OF THE MATERIALS EMPLOYED IN CARRYING IT OUT, AND SOME APPLICATIONS OF THE METHOD TO OPERATIONS AND INJURIES IN DIFFERENT REGIONS OF THE BODY, AND TO WOUNDS RECEIVED IN WAR

BY

WILLIAM MAC CORMAC, M.A. F.R.C.S.E. & I.
M.CH. HON. CAUS.

SURGEON, AND LECTURER ON SURGERY, ST THOMAS'S HOSPITAL
CONSULTING SURGEON TO THE FRENCH HOSPITAL

LONDON
SMITH, ELDER, & CO., 15 WATERLOO PLACE
1880

PREFACE.

THE ANTISEPTIC METHOD initiated by Professor Lister fifteen years ago occupies at present so prominent a position in surgical practice as to compel the attention of all practical surgeons; and the time has surely arrived when its pretensions, and the theory on which these pretensions are founded, may claim a definitive expression of opinion. Foreign surgeons, especially the German ones, will say that this is already done; but in England there is by no means the same unanimity on the subject.

It was with an object of this kind I accepted the request of the Hon. Secretary of the South London Branch of the British Medical Association, to prepare a paper on the subject. Because of the limited time, only a portion of the address could be delivered; but it was published in full in the 'British Medical Journal' two days afterwards, and the object for which it was written was accomplished by the interesting debate which followed, in which so many surgeons of eminence took part.

It did not occur to me to give either the address or the debate any further publicity. Mr. Spencer Wells, however, whose own contribution is of such great interest, suggested that it should be done, and that in the interest of Surgical History a record in a separate form of the various facts and opinions expressed by the different speakers would be valuable.

All having willingly consented, this republication is now made. Errors—some of them inevitable, perhaps, in an imperfect shorthand report—have been corrected, so that a correct version is afforded of what each speaker intended and wished to say.

In the hope that I might increase the general interest in the publication, and make it useful to students and practitioners anxious to learn and practise the antiseptic method as it is carried out at the present time, I have ventured to add a short statement of the more important points connected with the theory on which the method is based, without an adequate comprehension of which I do not believe the antiseptic system can be successfully carried out.

In a second section the antiseptic materials in common use, and the manner of their preparation, are described.

A chapter is also devoted to some of the practical applications of the antiseptic method, including its employment in military surgery.

I do not profess to give a complete account, and there may be some repetition, as these chapters are partly covered by the matter of the address. This I have endeavoured to render more complete by giving the statistical portion in greater detail.

13 HARLEY STREET:
May 11, 1880.

CONTENTS.

	PAGE
INTRODUCTION	1
ADDRESS	1
DEBATE.	47
Mr. Bryant	47
Mr. Macnamara	49
Mr. Barwell	53
Mr. Spencer Wells	54
Mr. Thomas Smith	56
Mr. Timothy Holmes	57
Professor Lister, F.R.S.	59
ADJOURNED DEBATE.	
Professor John Wood, F.R.S.	71
Mr. Jonathan Hutchinson	77
Sir James Paget, LL.D., D.C.L., F.R.S.	81
Mr. Lund	88
Dr. Newman (Stamford)	89
Mr. Knowsley Thornton	91
Mr. Morrant Baker	94
Mr. Mac Cormac	96
THE ANTISEPTIC THEORY	100

	PAGE
ANTISEPTIC MATERIAL	116

Carbolic Acid—Carbolised gauze—Carbolic acid poisoning—Jute—Salicylic acid—Thymol—Boracic acid—Acetate of alumina—Chloride of zinc—Protective—Mackintosh—Bandages—Sutures and ligatures—Drainage tubes—Sponges—Spray producers.

ANTISEPTIC PRACTICE	162

Objects—Necessaries for—Mode of dressing—Abscess—Compound fracture—Osteotomy—Fractured patella—Ununited fracture—Puncture of joints—Abdominal section—Fæcal fistula—Radical cure of hernia—Ovariotomy—Hydatid of the liver—Injuries to the head—Ligature of blood-vessels—Empyema—Amputation of the breast—Atresia vaginæ et hymenalis—The use of the catheter—Radical cure of hydrocele—The antiseptic removal of cancer of the rectum—Removal of a tumour from the wall of the chest—Union by suture of divided nerves and tendons in open wounds—The management of wounds in which septic changes have taken place—The use of antiseptics in military surgery—Metric table.

INDEX	283

LIST OF ILLUSTRATIONS.

	PAGE
Fig. 1. Case of extreme genu valgum, before operation	11
„ 2. The same case, after operation	12
„ 3. Bullet-wound of head of tibia	39
„ 4. Küster's antiseptic gauze fabricator	121
„ 5. Bottle for preserving catgut and silk	138
„ 6. Lister's catgut-holder	139
„ 7. Trough for preserving catgut ligatures	139
„ 8. Esmarch's and Chiene's method of filo-pressure	140
„ 9. The leaden-plate suture	143
„ 10. Thiersch's bead suture	144
„ 11. Wills' suture buttons	145
„ 12. Drainage-tube prepared for insertion	146
„ 13. Sizes of drainage-tubing	147
„ 14. Lister's forceps for introducing drainage-tubes	148
„ 15. Hand spray-producer	153
„ 16. Richardson's spray, with foot bellows	154
„ 17. Steam spray	156
„ 18. Steam spray with spherical boiler	158
„ 19. Application of artery forceps in an amputation of thigh	165
„ 20. Clamp artery forceps	166
„ 21. Mode of introduction of deep and superficial sutures	168
„ 22. Amputation stump ready for dressing	169
„ 23. Wound syringe	173
„ 24. Antiseptic dressing for a wound in the groin	179

a

LIST OF ILLUSTRATIONS.

		PAGE
Fig. 25.	Plaster-of-Paris splint, with strengthening strips	184
,, 26.	Shape of anterior and posterior flannel leg splint.	185
,, 27.	Anterior leg splint, moulded	185
,, 28.	Posterior leg splint, moulded	186
,, 29.	Limb in plaster splints, suspended	186
,, 30.	Posterior plaster-of-Paris splint	187
,, 31.	Dressing for compound fracture of leg	187
,, 32.	Walker's roller for wet plaster bandages	188
,, 33.	Normal femur, and femur in genu valgum	190
,, 34.	Section of normal femur, and femur in genu valgum	191
,, 35.	Diagram showing relation of parts in genu valgum before and after operation	192
,, 36.	Vertical section of knee-joint illustrating the operation for ununited fracture of patella	194
,, 37.	Mason's suture for an ununited fracture	198
,, 38.	Ununited fracture of femur	198
,, 39.	Lembert's intestinal suture	207
,, 40.	Antiseptic dressing after ovariotomy	216
,, 41.	Temperature chart after a head injury	225
,, 42.	Temperature chart after a head injury	227
,, 43.	Comminuted portions of skull removed in a case of depressed fracture	228
,, 44.	Portions of skull removed in a case of depressed fracture, and their position	229
,, 45.	An antiseptic dressing for the head	231
,, 46.	Process of obliteration of an artery and the organisation of an animal ligature applied to it	233
,, 47.	Drainage-tubes for the pleural cavity	235
,, 48.	Antiseptic dressing as applied after amputation of the breast	237
,, 49.	Antiseptic dressing for excision of the shoulder-joint	238
,, 50.	Temperature chart showing aseptic temperatures in a case of operation for imperforate hymen	240
,, 51.	Pelvic support for use during dressings of the thigh, perineum, groin, scrotum, or rectum	250

LIST OF ILLUSTRATIONS. xi

	PAGE
Fig. 52. Drainage-tube for the peritoneal cavity	250
„ 53. Schucking's glass drop tube	251
„ 54. Large enchondroma growing from the ribs	253
„ 55. Appearance of chest-wall after removal	254
„ 56. (a) Sharp spoon. (b) Bone gouge	261
„ 57. Volkmann's splint for suspending the arm	264
„ 58. Apparatus for continuous irrigation	265
„ 59. Antiseptic tampon for gun-shot wounds	273
„ 60. Esmarch's first dressing for the wounded in battle. 1. Packet folded up. 2. Triangular bandage. 3. Gauze bandage. 4. Antiseptic tampon. 5. Tampon and square of oiled paper	274

ANTISEPTIC SURGERY.

A LARGELY ATTENDED and influential meeting of the South London Division of the Metropolitan Counties Branch of the British Medical Association was held at St. Thomas's Hospital, on Wednesday night, December 3, 1879, for the discussion of the subject of Antiseptic Surgery.

Mr. JOHN WOOD, F.R.S., a President of the Branch, occupied the chair; and, in opening the proceedings, said the subject they were invited to discuss was one which he thought fell especially within the scope of the Metropolitan Counties Branch of the Association, embracing, as this did, all the chief centres of medical and surgical thought in the metropolis. It seemed to him peculiarly appropriate that they should have the united opinions of the different schools upon a subject of such paramount interest as that of Antiseptic Surgery—a question so momentous that it would be difficult to exaggerate its pressing importance, not only to the Profession, but even still more to the public.

Mr. WILLIAM MAC CORMAC then addressed the meeting to the following effect :—

ADDRESS.

I suppose it may be assumed that all surgeons employ antiseptics, in some form or other, and I think their use, which, during the last fifteen years, has become more and more general, is very largely due to the attention drawn to the subject by the particular method of employing antiseptics with which the name of Professor Lister is associated. It would be impossible within any moderate limits to discuss the different ways in which antiseptics are used in the treatment of wounds. I therefore propose to confine myself to Lister's plan of employing antiseptics, including any modifications involving the same theory and principles. Other methods I would term inexact. The Listerian is, in contradistinction, an exact method, founded on a special theory, and carried out in every detail in almost precisely the same manner in each case. The theory is simple. Pasteur's beautiful experiments, which have never been disproved, show that it is not the air, but something contained in the air, which excites decomposition. The active agents concerned are considered to be certain minute vegetable organisms and their germs, which exist in various forms, and probably act in many different ways, and are all capable of being developed *ad infinitum* wherever they discover a suitable nidus. What the essence of the mischief consists in, or what it is precisely which produces the poisonous changes in the organism, need not now be discussed. One great fact is certain, namely, that the free access of ordinary air to a wound is very generally the cause of inflammatory or putrefactive changes in that wound; while experiment proves, and clinical experience corroborates the fact, that air which has been filtered, overheated, or subjected to the influence of certain germicides, like carbolic acid, is not capable of exciting inflammation or putrefaction in animal tissue or wound-exudations, and is harmless in its action upon wounds. Both observation and experiment establish, I think, that, in the great majority of cases, the disorders included under the comprehensive term of 'blood-poisoning' depend upon putrefying or septic fluids in a wound gaining admission to the general circulation. In short, if only wound-putrefaction be prevented, we eliminate from the deaths occurring after surgical operations or injury those which are produced by blood-poisoning in its various forms—pyæmia, septicæmia, and, in a lesser degree erysipelas. That blood-poisoning may occur independently of external wound must be conceded. This is, however, exceptional. As a broad clinical fact, I think it may be assumed that, in surgical practice, blood-poisoning, arising from altered secretions in the wound, has hitherto been the ever-present dread in the minds of those who are dealing with wounds, whether inflicted by the knife or arising from accidental violence. This is the main and all-important difference between subcutaneous injury and an open wound. A simple fracture, if kept properly adjusted, unites easily and quickly, with scarcely any pain or fever, and almost no risk; whereas a quite similar injury of a bone, if associated with external wound, excites fever, and there is pain, with possibly extensive suppuration or necrosis. The limb may be eventually sacrificed, or, in a proportion of the cases, the sufferer will lose his life. Now, if it be true that these formidable consequences owe their origin to putrefaction of the wound-secretions, that putrefaction depends upon the free admission to the wound of minute living particles in the atmosphere, and that by the methodical use of certain germicide agents, these putrefactive changes in wounds and their consequences may be prevented absolutely, then the various forms of blood-poisoning may be erased from the list of surgical diseases.

In addition to this, the advocates of antiseptics allege that by their method, the duration of treatment in cases of injury or operation is, in the main, curtailed; that the safe and painless healing of wounds is promoted, the general comfort and well-being of the patient increased; that many operations, too dangerous to be performed, except of absolute necessity, may, with the safeguards now at our disposal, be undertaken without risk; that operations of expediency—that is, operations for the removal of deformity, or desirable for other reasons, hitherto considered unjustifiable on account of the risk, may be performed with so much safety as to become matters of almost routine occurrence. It is not denied that, under what may be called the less exact antiseptic methods of treatment, inflammation and putrefaction may be slight or altogether absent; that wounds may unite by the first intention, patients recover well, and operations of various kinds prove successful—the human body being fortunately capable of resisting evil influences within certain limits; but what is contended is, that this result cannot be counted upon as matter of course; whereas, with the exact antiseptic method, wound-putrefaction and all its consequences can be prevented with a certainty unknown in any other way. That cases of failure have from time to time occurred is undoubted; but those who are convinced of the efficacy of Lister's method believe these failures arise from some imperfection in the manner of using it; some unconscious neglect of those details, a punctilious attention to which is so all-essential.

Professor Lister's method, from time to time variously modified, in his own hands has now reached so high a degree of perfection as to be the one generally employed. It has not, I think, thus far been improved upon by others in any essential feature.

It will now be convenient to state what materials are used, and the manner of using them. If an ordinary operation is to be performed, the surface adjacent, if covered with hair, must be shaved, then thoroughly washed with a five-per-cent. solution of carbolic acid. The actual steps of the operation are conducted in a carbolised atmosphere, produced by a jet of steam mingling with a five-per-cent. solution of the acid. The sponges employed, the hands of the operator and those of his assistants are thoroughly purified in a five-per-cent. solution, previously to the operation, and again and again during its progress. The instruments are kept ready in a two-and-a-half or three-per-cent. solution, which may be also used for washing the wound and the sponges. All bleeding points must be carefully secured, either by torsion, carbolised gut, or carbolised silk, the ends being cut short. Too much time can scarcely be devoted to the complete arrest of hæmorrhage. The sutures should be both deep and superficial; the former of wire, the latter of catgut, and if there be tension, the button-suture is useful —a simple form consisting of leaden plates, through which the wire passes. The entire surface of the wound should, if possible, be brought into apposition. Drainage-tubes should be inserted, in order that the bloody serum which exudes may easily escape externally, without producing tension or separation of the wound-surfaces, as well as because of its proneness to decompose. The principle on which these act is to carry the secretion from the deeper parts of the wound, as well as from any irregularity or recess, by the straightest shortest road to the surface. Therefore, they ought to be of sufficiently large calibre. Many short drains, rather than few and long ones, are preferable. They should be removed so soon as their function is at an end—in many cases, at the end of three or four days, and always before they set up irritation, or act as foreign bodies. Thus used, drainage is of the first importance : it is primary and preventive. The drainage-tubes of Chassaignac served a different object : they were intended to provide for the escape of matter already formed, and might be called secondary drains. A sufficient number of tubes having been inserted, the projecting portions are cut off level with the surface, and a layer of protective silk applied to the wound. Over this are placed several layers of carbolised gauze, wrung as dry as possible out of a two-and-a-half-per-cent. solution of carbolic acid, and fastened to the surface with a carbolised bandage. Over this again may be applied a layer of salicylic wool. It exercises an elastic equable compression, and fills up any irregularities of the surface, besides forming an antiseptic filter at the margins of the dry gauze, now applied as an eight-fold layer over all, and secured by a bandage—a piece of macintosh being interposed between the last layers of the gauze. The dressing should extend two or three, or several hands' breadth, according to circumstances, beyond the extremities of the wound; and in places where there is much body movement it is convenient to keep the margins of the dressing closely applied to the surface by a turn of an elastic web bandage. Many parts of the body present special difficulties, as the neck and groin. In all cases the first dressing is the most important. Too much time and care cannot be devoted to it, as on its success depends the whole future of the case. At a first dressing too much carbolic acid cannot well be used, while at subsequent dressings the less the carbolic acid comes in contact with the wound the better. The ideal wound-dressing, in short, is that which affords perfect rest, free exit for secretions, an absence of tension, with antiseptic protection. In contused or lacerated wounds, the stage of 'cleaning' and accompanying discharge are reduced by this treatment to a minimum, or do not take place at all. Incised wounds or amputations heal more frequently and readily by the first intention, without redness, swelling, or pain; while blood-clots, if undisturbed, remain dark-coloured and without change, until granulation-tissue enters into and occupies them. Septic suppuration and necrosial cellulitis do not occur. The early union in an amputation tends to avert those disadvantages which may follow protracted healing, the retraction of the soft parts, or the adhesion of the cicatrix to the bone. Pain, fever, and duration of treatment are diminished.

A simple plan for the manufacture of carbolic gauze, devised by Professor Bruns, of Tübingen, seems worthy of trial. To prepare one kilogramme of this gauze, four hundred parts of resin and two litres of spirit are required. After being completely dissolved, one hundred parts of carbolic acid are added, and eighty of castor-oil or one hundred of stearine. After soaking, the gauze is hung up, to allow the spirit to escape. A few minutes suffice to dry it, and then it is packed in air-tight cases. It can be freshly made with great ease. It contains eight or nine per cent. of carbolic acid, while the ordinary gauze contains six or seven. It is stated to be more pliable, less irritating, and cheaper than the ordinary kind. Bruns has prepared an extract, two litres of which suffice for one hundred *mètres* of gauze. This is very portable, and applicable for use in military surgery or in country practice.

Quite lately Dr. Neuber of Kiel has used drains made of decalcified bone, with these advantages :—that, after having served their temporary purpose, the drains melt away, the dressings do not require to be changed for their removal, and very often it happens that a dressing would not be disturbed, except indeed to see after the drainage tubes. These drains are much less likely to irritate than india-rubber tubes, are easily prepared by soaking tubes, prepared from healthy ox or horse bone, in a mixture of one part of hydrochloric acid in two of water, for ten hours, and are preserved for use in ten-per-cent. carbolic oil, after first washing them in a five-per-cent. watery solution. Thus treated, wounds may remain, Dr. Neuber finds, without change of dressing from two to four weeks. In large operations, or those where any bleeding may be expected, Neuber applies a second dressing twenty-four hours after the first. He states that the cases in Kiel recover without pain, fever, or suppuration, and that the wound-surfaces in the greater number of instances immediately glue together.

Professor Maas of Freiburg, chiefly on the score of the

poisonous effects of carbolic acid, and its volatile properties, has employed a solution of the acetate of alumina, which may be prepared in a strength of 15 per cent. by adding dilute acetic acid to hydrate of alumina. A three-per-cent. solution arrests the development of bacteria, while four or five per cent. of carbolic acid is required for the purpose. Compresses soaked in a two-and-a-half-per-cent. solution of the alumina are applied, under a spray of the same strength, and a number of amputations, resections, and washings out of the larger joints, treated in this way, succeeded admirably in the Klinik at Freiburg. There was very little secretion from the wounds. As an antiseptic, it is perfectly safe, and the cost is inconsiderable. It can, however, only be used in a moist form.

After a long trial of thymol, I found it too feeble an antiseptic to be safely employed in any except quite small operations, or when but little secretion was anticipated, while it possesses no particular advantage over carbol. I place but little reliance upon thymol spray, which is only one part of thymol in 2000. It gave me the impression that it differed but little from simple steam.

The rough, disagreeable condition of the skin of the hands, which first becomes very dirty in appearance, and then peels off, caused by immersion in carbolic solution or spray, may be considerably abated by protecting the surface with carbolic vaseline, one part in 10; but it has the disadvantage of making the grasp slippery.

The employment of Lister's method is not at once nor very easily acquired; it requires practice, together with a patience and capacity for detail. It requires, also, that those concerned in the management of a case shall, without reserve, believe in the germ theory, or act as if they believed it. The surgeon in charge of the case must either himself examine into and verify everything belonging to the dressing of that case, or have some one in whom he can thoroughly trust to do this for him. Less than this will not be putting Lister's method fairly to the test. This is a difficulty in hospital practice, where it is impossible to superintend all the details yourself, and where your house-surgeons or dressers may not be sufficiently impressed with the necessity of a minute attention to details. In this respect the German professors have an advantage over us. In every German hospital there are one chief surgeon, and three assistants who hold office for three years, and who are generally much senior to our house-surgeons, and the management of the whole surgical practice of the hospital is under the control of one man.

Statistical Results.—When, for the purpose of this discussion, I began to seek for subject-matter in the shape of statistical facts, I experienced some difficulty in obtaining what I wanted. In our own country, the published record consists chiefly of isolated cases, or of groups of a few cases, and of individual impressions; but one important deduction may be arrived at, that, with few exceptions, those who have practised Lister's method find it a good method, and admit, in a greater or less degree, the justice of the claims put forth on its behalf. This and not mere statistics, which vary enormously in value, seems to me an important point. I come myself to a conclusion—and others do the same—that a certain method of treatment is the best, because I find my results are better than those I previously had; because cases, about which I should before have had the greatest anxiety, now occasion me none; because I can perform operations on which I should not otherwise have ventured, with complete confidence beforehand in the result; and because my patients appear to recover more quickly, more easily, as well as more safely. There is another important factor, which is, that the class of cases now submitted to surgical treatment has, to some extent, changed. Operations are now daily and in large numbers practised, which were never, or only rarely, before attempted. With these qualifications, let us examine some statistics. They are only meant as a contribution, and one far from being so complete as I could have wished.

In private practice, the cases are too few to afford means of obtaining a statistical comparison, but in cases of amputation, or removal of tumours, where I have been able to apply the antiseptic method thoroughly, the results have equalled my most sanguine expectations. I have been able to form a strong impression of the value of the method from seeing wounds heal under it, by immediate union, without a trace of local redness, swelling, or even pain, without suppuration generally, and often without constitutional reaction in the slightest degree. These are isolated cases, but they acquire a legitimate value when, in considering the merits of this question, one may predict beforehand that, under such and such conditions, certain results shall follow.

With regard to our own hospital statistics, I am not able to place before you, perhaps, a large array; but I present some which, so far as they extend, are not wanting in importance. In the first place, we have had, during the last eighteen months, 45 operations for the division of various bones, in order to remove deformity.[1] This division has been accomplished in some cases with the chisel, and in some with the saw. In the majority—30 in number—the knee-joint was engaged in the operation; in some, the shafts of the long bones were chiselled through; and in others, a wedge-shaped piece of bone was removed. The wound in the soft parts was comparatively small, but quite large enough in many instances to allow serious inflammatory and putrefac-

---

[1] Since this Address was written 12 additional cases of osteotomy have been treated with the like result of complete recovery, without any unpleasant symptom whatever.

tive changes to take place in it; yet not only in no single instance, among all these cases, did a serious result of any kind follow, but there was a complete and speedy recovery, without more inconvenience, indeed, than is entailed by keeping a limb in plaster-of-Paris splints for simple fracture. The point is, that in every one of this considerable series recovery ensued in the speediest, easiest, and safest manner,

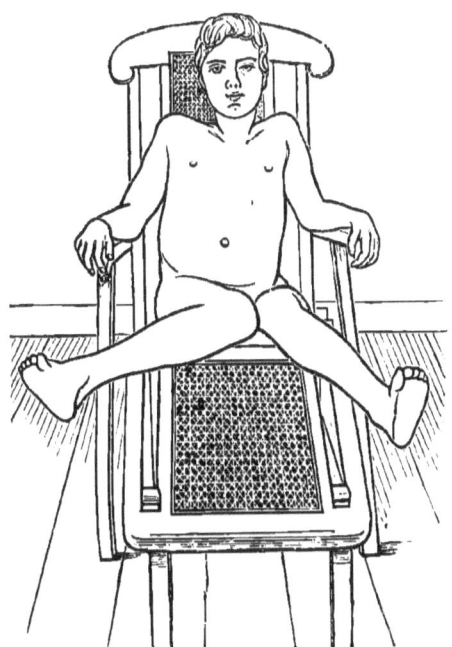

FIG. 1.—R.D. Extreme genu valgum, taken from a photograph. The right leg was at a right angle with the thigh when the limb was extended.

and in every one in which the knee-joint was opened—some 30 joints in all—a freely movable articulation resulted.

I may mention one of the cases in illustration—namely, that of a boy, æt. 14, with extreme genu valgum (fig. 1). The left limb I straightened by means of Ogston's division of the internal condyle. In the right, the deformity was more pronounced; and this operation only succeeded in very partially removing it. I then divided in succession, on the outer side of the limb, the biceps tendon, the ilio-tibial band, and the external lateral ligament. Air freely entered the joint, already filled with blood and bone-dust. I used a great

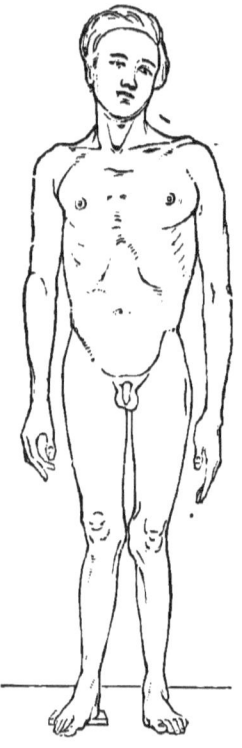

Fig. 2.—R.D. Result of the operations. The right upper and lower limbs are somewhat shorter than those of the opposite side. The boy is extraordinarily developed in the upper half of his body.

deal of force in my efforts to straighten the limb, but, so great was the previous deformity, without much avail. However, after all this violence done to the greatest articulation in the body, the patient recovered without any ill result, even in the most trifling degree. Subsequently I divided the right femur just above the condyles, and the boy now possesses two almost straight limbs and two good movable knee-joints. The woodcuts (figs. 1 and 2) represent his condition before and after operation.

In regard to *compound fractures*, our antiseptic practice is not yet very extensive; but it has proved, I think, eminently satisfactory. Sixteen cases of compound fracture—the whole number, excluding those complicated with wound of joint, or requiring immediate amputation, or dying immediately— have been treated on strict antiseptic principles during the last eighteen months. Two cases were of the femur, 3 of the upper extremity, and 11 of the tibia—all did well. In 1 case, the dressings became offensive, and amputation was performed. For some reason, they had been daily changed in this case; while now a change is made twice or thrice altogether. In one case of fracture of the femur, union has not yet taken place, without assignable cause for the delay. All the cases of fracture of the leg recovered almost as if they had been simple fractures. On examination of the reports for the last six years, Mr. Pitts, who was good enough to make these enquiries, finds there have been 54 cases of compound fracture of the lower extremity not treated antiseptically. From this list, also, have been excluded cases involving injury to the joints, and those requiring immediate amputation. Among these 54 cases were twelve deaths, 4 of them being from pyæmia and 4 from erysipelas. Thus, 8 deaths are due to preventable causes. Of the 42 cases that recovered, in one, secondary amputation was performed for gangrene of foot. In 25 cases recovery took place without complication or delay. Of the remaining 16 cases, 4 had erysipelas, but recovered after a long illness of from 77 to 211 days; in 3, there was delayed union on account of necrosed bone, recovery taking place after 144, 105, and 209 days respectively. In the remaining 9 cases there was suppuration and a more protracted recovery than with any of the antiseptic cases. On examination, we find that the antiseptic cases, on an average, remained under treatment in hospital for a period of little more than two-thirds the length of time of the others. Most of the cases which recovered quickly, and were not treated antiseptically, were cases treated by sealing, the fracture being caused by indirect violence, the wound small, and generally occasioned by the sharp end of an oblique fragment coming through the skin. In nearly every other instance of compound fracture with much laceration of soft parts, whether treated by sealing the wound, by covering it with oiled lint, or the application of wet lint, &c.—recovery has only taken place after a long stay in hospital. Protracted suppuration, necessitating frequent counteropenings, being the rule, the patient not unfrequently was attacked by erysipelas, consequent, perhaps, in many cases, on the frequent handling and exposure of the wound.[1]

In the last volume of 'St. George's Hospital Reports,' Vol. IX., Mr. Holmes has published a paper on compound

---

[1] From the cases treated antiseptically we find, according to the report Mr. Pitts has made up to the present date—

Femur, 2 cases. In both no suppuration occurred nor any rise of temperature.

Tibia, or Tibia and Fibula, 19 cases. Fourteen healed as simple fractures, and did not require any after counteropenings or incisions. No rise of temperature or necessity for redressing more than once for the primary injury. In two cases protracted recovery. In one of these two inches of projecting tibia had to be sawn off—there was much laceration of soft parts, and but for the youth of the patient a primary amputation would have been resorted to. In the other case of protracted recovery, a large portion of skin sloughed. There was great bruising of the whole leg, caused by fall of a large piece of iron. Bone was exposed for several inches—but is now covered by granulation. Patient still under treatment. In 2 cases, antiseptic treatment failed to prevent diffuse suppuration, and secondary amputation had to be performed. In one case, patient died after nearly seven weeks in Hospital. This was a very bad case of compound fracture of leg in an old man, complicated by an injury to the head. Patient died from an abscess in brain, quite independent of his injury to leg. The antiseptic treatment of his compound fracture was completely successful. Of the upper extremity, 3 cases occurred, all successful.

Cases of compound fracture of lower extremity not treated antiseptically. Excluding cases involving injury to joints, and those requiring primary ampu-

fractures of the leg of much interest. It may be taken as fairly representing what has hitherto been the ordinary rate of mortality after compound fractures in a London hospital when antiseptic practice has not been carried out.

From 1865 to 1878, 162 cases of compound fracture of the leg were treated to a conclusion. From this number are excluded 31 cases requiring immediate amputation, also 10 cases which died within 48 hours of admission. Of the 162 cases, 40 or nearly one-fourth died. They were treated without antiseptics, and 21 or rather more than one-half of the fatal cases died from pyæmia. Phagedæna occurred in four cases, diffuse cellulitis in six, erysipelas in six, tetanus in one.

The number of complete antiseptic cases treated in St. George's Hospital is reported as 33, of which 7 died. In three of the fatal cases, the patient had delirium tremens, one of the three dying eventually from pyæmia. In a fourth gangrene supervened. In a fifth death was caused by tetanus. The sixth had visceral disease; and the seventh patient, who died from pyæmia after secondary amputation, is 'the only one,' Mr. Holmes says, 'in which the

---

tation—during the years 1874-79. Number of cases, 54. Deaths, 12, viz.: Pyæmia 4, Erysipelas 4, Delirium Tremens and old age 1, Tetanus 1, Old age, 76, and shock 1, Suppuration in leg, and exhaustion from Phthisis 1. Thus, 8 deaths appear to be due to preventable causes amongst the 54 cases.

Taking the 42 cases that recovered we find—In one, secondary amputation of leg was performed for gangrene of foot. In 25 cases recovery took place without any complication or delay. Of the remaining 16 cases, 4 had erysipelas, but recovered after a long illness—viz. of 211, 77, 174, and 195 days respectively. In 3 there was delayed union, on account of necrosed bone, with recovery after 144, 105, and 209 days respectively. In the remaining 9 cases there was suppuration, and also a protracted recovery. In one case the patient had to remain in Hospital for 160 days.

The average duration of stay in Hospital of the cases of compound fracture of lower limb treated in the ordinary way was a little over 80 days.

Average duration of cases of same class treated antiseptically is found to be about 60 days. Cases still in the Hospital are necessarily excluded from this average. The majority of the injuries were of a severe character.

antiseptic treatment can be said to have been followed by blood-poisoning.' This is surely greatly different from the previously recorded experience, where one half of the large percentage of fatal cases died from pyæmia. Yet still the percentage of death is large, and compares, I think, unfavourably with such a series as that published by Dr. Wildt from the practice of the Neues Städtisches Krankenhaus in Berlin, not to speak of Volkmann's series. In 28 cases 2 deaths occurred, one from tetanus, the other from septicæmia.[1]

Mr. Holmes says: 'Very exaggerated estimates have been formed of the difference which the introduction of antiseptic surgery is believed to have made in the treatment of ordinary cases. These estimates are mainly based on statistics published by foreign surgeons from hospitals where little atten-

---

[1] H. Wildt, '*Die Complicirten Fracturen,*' *Centralblatt* 47 und 48, 1877.
Tabular arrangement of 28 compound fractures treated in the Neues Städtisches Krankenhaus, Berlin. Amongst them are—

Arm 2
Forearm 7
Thigh 2
Leg 17

Nineteen were caused by direct violence.

In 21 cases the fractures were brought to the Hospital during the first 24 hours, and before any important reaction had occurred in the neighbourhood of the fracture.

These all recovered. Two intermediary amputations were, however, necessary for gangrene. The other 19 almost entirely healed without suppuration beneath 'the soft blood clot scab.' In the three which did not progress so favourably, in one the wound became stinking, with the approach of gangrene, and in another it became diphtheritic.

Seven cases were admitted later than 24 hours after the injury, with more or less manifest reaction in the neighbourhood of the wound.

Two of these were very severe injuries, and ended fatally: one, a bad crush of hand and wrist, by tetanus on the ninth day. The other case died of septicæmia dependent on erysipelas and wound-diphtheria.

Of the other five, although the foul-smelling wounds were soon made aseptic, in no case did healing under the moist scab take place. In one case amputation was necessitated by gangrene of the arm following ligature of the brachial artery for hæmorrhage.

tion seems previously to have been paid to hospital management, and where consequently pyæmia and erysipelas were allowed to prevail unchecked.'

I am not concerned with the latter part of this statement further than to point out that whatever gain in human life and saving of human suffering these 'foreign surgeons' have accomplished, have been realised under conditions which are said to be, and with some truth perhaps, opposed to the successful attainment of these objects.

The same thing has been stated before in other words, and it amounts to this: that the former results were very bad compared with those obtained elsewhere, which I think is true, and the admitted improvement is merely a relative one, that absolutely the results are not now at all so good as those to which we are accustomed, and the effect, if not the intention, of the argument has been to depreciate the value of Lister's treatment.

It is difficult to believe that there is a conspiracy amongst the German surgeons to suppress the truth, for it is to German surgeons that the observations I have alluded to have been, I think, directed.

We must regard these men as honourable fellow-workers in the paths of science. I am happy to count many amongst them as my personal friends, and I know and I hope appreciate their enthusiasm and energy in the pursuit of surgical pathology and practice.

Well, Professor Volkmann has had in succession 75 cases of compound fracture without a death, which will compare favourably with surgical practice in any part of the world. If this be not a success, I do not know what success is— and the major part of this success was accomplished in one of the oldest, most unhealthy, and unsuitable hospital buildings on the continent of Europe.

The way in which ordinary cases of operation recover

seems much easier under the antiseptic method. For instance, in an amputation at the hip-joint in a young man, for a large sarcoma of the thigh, the patient recovered rapidly, without giving a moment's anxiety. His previously hectic state due to the rapid growth of the tumour almost at once subsided. There was immediate union of the flaps of the wound, save where the drains emerged. The deep catgut stitches, inserted to draw the surfaces together, disappeared; the dried-up knots were picked off the surface of the stump. In a fortnight, the patient was sitting in a chair, quite convalescent.

Recently a patient has been under my care for chronic synovitis of the knee-joint, which had resisted various modes of treatment for seven years past.

The case seemed one of Hydrops Articuli; the patient was young, and his health otherwise good. I cut into the joint, and after the evacuation of the glutinous synovial fluid, which contained a few flakes of lymph, washed it out several times completely, with five-per-cent carbolic solution, and inserted a drainage tube, as large as my little finger, into the joint. Contrary to my intention and wish, the house-surgeon washed out the joint again next day, and possibly thus excited some irritation, for a localised abscess formed near the drainage tube. But the suppuration was aseptic; the fluid has not reaccumulated; the joint is now two inches smaller in circumference than before, and the same girth as the other knee, the man is able to walk about with a fairly movable joint, and is otherwise quite well.

Three weeks ago, I performed *Syme's amputation* in a feeble lady, aged 63, who had suffered for ten years from ankle-disease—*tumor albus.* I first scraped away the granulation-tissue from the sinuses, disinfecting the neighbouring parts beforehand by soaking them in five-per-cent. solution of carbolic acid. The os calcis was much atrophied;

it broke in two during the removal of the foot. I most carefully scraped or dissected away all diseased structures. The patient recovered without an atom of fever, pain, suppuration, or even discomfort. In fact, her well-being improved from the operation-day. The external wound united by first intention, except at the two drainage openings, without a trace of local inflammation. Five dressings were made, the last two at intervals of a week. At the end of three weeks, a small superficial granulating surface might be seen at the point of emergence of the drains, and at one of the larger of the old sinuses; the others have closed. The patient can bear any ordinary pressure on the end of the stump. This was an absolutely good result under what appeared to be unfavourable circumstances.

In a case of *movable body in the knee-joint* in a young man, I lately cut down and removed a body, the size of a broad bean, through a corresponding opening in the soft parts, at the same time allowing a considerable quantity of synovial fluid to escape, the result of a preceding attack of acute synovitis. The man was relieved of his loose cartilage and the synovitis at once, without a trace of local or general disturbance. The wound into the joint was quite an inch long, and I was in no hurry to close it on account of some bleeding. Yet I had no anxiety about the result. A free incision into the knee-joint without antiseptic precautions cannot be made without causing great anxiety. I have seen a strong, healthy man lose first his limb, and afterwards his life, from acute septic suppuration arising in the joint after such an operation. Larrey collected 135 cases of the direct method of operation, with 30 deaths, or 22 per cent.; and many fatal cases which have occurred have not been published. The indirect operation is comparatively quite safe, but is attended by a large proportion of failures.[1]

---

[1] On looking over the records of cases of removal of loose cartilage from

The total number of antiseptic cases of *ovariotomy* performed in the Hospital has been 19, amongst which 7 deaths occurred. The fatal cases were very severe; in one, the bladder was adherent to the abdominal wall and opened into; in one the operation could not be finished; in another the base of the uterus had to be cut away; in none, however, the surgical registrar informs me, did death occur from septic poisoning. In the preceding three years, it happened that the same number of cases were also operated on—namely, 19, amongst which 13 proved fatal.

I thought it would be interesting to collect the experience of ovariotomy, performed before and after antiseptic treatment, and as a sort of test-operation.

Mr. Spencer Wells has been good enough to tell me that he operated in 83 cases antiseptically, with but 6 deaths, the last 38 in succession recovering. On one occasion, before the use of antiseptics, he had a run of 27 successful cases, and twice of 21. Of the 6 cases which died under

the joints, the difference in the safety of the subcutaneous method and that by free incision in the joint is very apparent.

Mr. Square of Plymouth has had a series of twenty-four operations performed in succession with complete success, but this is an almost unique experience.[1]

Benndorf, Inauguration Thesis, quoted in Schmidt's *Jahrbuch*, 1868, p. 313, analyses a series of cases with the following results for the two methods:—

	Direct Incision	Indirect Incision
Cure	81 per cent.	66 per cent.
Deaths	19 ,,	10 ,,
Failures	0·0 ,,	24 ,,

The large number of failures to extract the body on account of its getting lost in the joint, induced many surgeons to prefer the direct incision, with its greater risks, and what these were may be seen in the cases recorded of protracted suppuration followed by anchylosis of the joint, or the necessity for amputation, or by the actual loss of life itself.

The Antiseptic method permits the direct method of removal to be practised with safety, and the risks of failure which the surgeon was exposed to in the subcutaneous method, although these may be greatly lessened by following Mr. Square's directions, are practically set aside.

---

[1] *British Medical Journal*, September 23, 1871.

antiseptic treatment, there was only one case in which there was any doubt as to the asepticity, the others were certainly aseptic. It has been said the practice of ovariotomy has been so successful without antiseptics, that the percentage of improvement proves little. But this is not so. Mr. Wells and others have had runs of unsuccessful cases of 7 and 9 in number in succession, and the question is, how many of these deaths, nearly all of which were probably septic, might have been averted.

Mr. Knowsley Thornton writes, October 23rd, 'Antiseptic treatment has exactly reduced my ovariotomy mortality by half, and would have done much more than this had I not lost cases through their coming to me with putrid cysts, the result of tapping without antiseptics. My mortality would have been lower also, had I not considered it a duty to give every case the chance of life afforded by operation, however hopeless it might appear. Since I began ovariotomy, I have only refused two cases, both clearly malignant. Freedom from fever and other discomforts, and more rapid convalescence, are the benefits derived by the patients who recover; immense difference in after anxiety, dressings, &c., for the surgeon.'

Dr. Keith of Edinburgh writes to me, under date of November 9th, that he has had altogether over 305 ovariotomies, which, arranged in fifties, show the following mortality:

	Deaths
First fifty	11
Second fifty	8
Third fifty	8
Fourth fifty	6
Fifth fifty	4
Sixth fifty	0

The last 76 of these cases were done antiseptically, with only 2 deaths. The last 68 cases in succession all recovered. 'I would give something to know,' Dr. Keith says, 'what the mortality of the last fifty would have been in the old way. To compare the cautery cases : there were 52 without antiseptics, and 4 deaths ; these were the worst cases, where, from the shortness or thickness of the pedicle, or other causes, the clamp could not be used. 55 cautery cases done antiseptically showed no deaths.'

Professor Nussbaum writes to me, November 11th, 1879, giving his experience of ovariotomy from February 1861 to October 1879. 'My first 5 cases were all fatal; of the first 34, 16 died ; of the first 78, 35 died. Then I employed drainage, and of 84 cases treated in this way, 34, or nearly one-half, died. Up to October 12th, 1879, I operated according to Lister's method in 135 cases, of which 29 died ; and in the last 19 cases there were 2 deaths. In my whole life, I have operated on 219 cases, with 67 deaths. I have never left a single operation, even in the worst cases, incomplete. These cases usually die in the hands of others, while some cases of this kind recovered in my hands. The general wards in the Munich Klinik, from being rife with pyæmia and gangrene, now never contain a case, either of one or the other. The mortality is reduced to one-half, and the only cases brought to the *post-mortem* room are those of deaths by suicide, from severe mechanical injury, in old people, or from cancer and tubercle.' Nussbaum fairly asserts that Professor von Buhl, the pathologist, his assistants, and the students of the hospital, would scarcely become partners in a wilful deception, when he says he now never sees a case of pyæmia.

Professor Esmarch writes to me from Kiel, November 10, 1879, that 'the practice of antiseptics is now so perfected, especially by the introduction of decalcified bone-drains, that most of our large operation-wounds, such as amputation of the thigh, extirpation of the mamma, with clearing out of the axilla, the extirpation of glandular masses from the neck, heal by the first intention, with a single dressing, without a trace of suppuration or wound-fever. Since antiseptics, our ovariotomies are incomparably better. In the previous year, I lost 1 case in 8 operations, and have once had 9 recoveries in succession—most of them very severe cases. My assistants have now very little to do during the visit, but the first dressing takes much time and requires much care.'

In his Hospital Report for 1878, Esmarch states that the antiseptic treatment is conducted in Kiel with increasing precision and increasing security. In 524 operations, large and small, the mortality was 25 in all. Of these 25 deaths, one was from exhaustion in an habitual drunkard, admitted for gangrene of the leg; 2 died on the day of admission; 1 died from pyæmia, after lithotomy; and 5 cases were in a septic condition on admission; 2 deaths occurred from cancerous cachexia; 3 from tubercle; 1 from delirium tremens; 1 from pyelo-nephritis; 1 hernia case died from peritonitis arising before admission; 1 patient died from secondary hæmorrhage after amputation at the shoulder-joint. 3 cases only appear to have died from blood-poisoning arising within the hospital, viz., 2 after amputation of the breast, and 1 after lithotomy; 1 ovariotomy case died from peritonitis. There were, therefore, 4 avoidable deaths amongst 524 operation cases of all kinds. Amongst the operations were 40 amputations, 27 resections of capital importance, 84 cases of removal of tumours, and 35 operations for necrosis.

In a further communication on the subject of infrequent dressing combined with the absorbable bone-drains, Dr. Neuber[1] gives good reasons for his satisfaction with the results of a more extended experience of their use in the hospital at Kiel. Indeed, it appears to me a discovery likely to prove only second in usefulness to that of catgut itself.

[1] Langenbeck's *Archiv*, vol. xxv. part 1.

In the majority of cases the granulations surround the tube, penetrate it, and finally replace it. Where the secretion is copious, which is unusual in aseptic wounds, the tube may soften and be discharged into the dressings. If completely surrounded by a coagulum, it remains without change for fourteen days or even longer. This, however, amongst 200 cases was only seen in a single instance. In contact with dead tissue the drainage tube remains unaltered. Tables of 131 cases of operation occurring from April to October 1879 are given, with full details. 3 were fatal, —one an amputation of thigh from septicæmia existing before admission; a second from pneumonia in a man of 70, from whom an immense tumour of neck had been excised; a third fatal result was in a weak child whose hip was excised; death occurring in 6 hours from collapse. None of these deaths could be properly connected in any way with the method of wound-treatment.

In 101 instances recovery took place after one dressing, generally without rise of temperature, and, except in 6 cases, without suppuration. The advantages of such a plan of treatment, and of having to dress cases of operation only once as a rule, are so evident as not to need enumeration.

In only 12 per cent. of these cases was there a rise of temperature, due to what Volkmann calls aseptic fever, that is a condition in which though the temperature be above normal, there is no other indication of constitutional disturbance, the patient appears to be quite well, and the progress and aseptic condition of the wound are not interfered with.

In the remaining 27 cases two or three dressings were required, but the greater part of the wounds healed in all by first intention. With the exception of the case of septicæmia mentioned, there was no wound-disease of any kind, nor was the aseptic progress of healing ever interfered with.

Such long-applied dressings have a peculiar cheesy smell

on removal from the accumulation, probably, of secretion from the skin glands. A few times an odour of decomposition was observed. But there was no bad consequence, as the wound had either healed or become a mere superficial granulating surface.

Drainage may be unnecessary in some cases, as small incised wounds, or those well covered with a flap of skin whose interior has not been irritated, and in which all bleeding is perfectly arrested. But in large wounds, in spite of our best efforts to stop bleeding and to avoid irritating the surface, drainage will be necessary, and a drainage tube which is absolutely unirritating, and which by the time its purposes are served disappears spontaneously without having produced any disturbance, is really a most considerable gain for the practice of antiseptic surgery.

Besides the mechanical advantages which drainage affords, there is also the very important prophylactic security given by it. Our method in some particular case may have been faulty, and septic particles may have gained admission. These, though perhaps unable to act injuriously upon the living tissue, can easily set up decomposition in the very unstable wound-secretion, if it be allowed to accumulate in the interior.

The method of dressing, as now carried out by Neuber, consists in applying a large, thick pad of carbolic jute enclosed in carbolic gauze, in place of the folds of loose gauze.

It is much cheaper, and exercises a gentle uniform compression on the wound.

The dressings are fastened on by india-rubber bandages.

Sixty additional cases occurred during the months from October to December, and 42 of these recovered after one dressing. They included cases of amputation, resection, removal of mamma and of other tumours, necrosis operations and compound fractures. There was no death, and no case of accidental wound-disease. Dr. Neuber offers this modification as a contribution to the simplification of Lister's method. The time of the surgeon is saved, the cost of material greatly lessened, while the advantage to the patient appears to be in no respect diminished.

In 1871, Volkmann's Klinik was so full, and the deaths from pyæmia and erysipelas so numerous, that he was on the point of closing it altogether. Some time after this, the antiseptic method was introduced. Volkmann states that the mortality from excision of the knee by German surgeons was not more than half the number of cases operated on before the employment of antiseptics. During the last three years, he has excised 21 knees, with a single fatal result from tuberculous meningitis, three weeks after operation. Two recovered after subsequent amputation. Recovery took place in the other cases very like that after simple osteotomy, generally with primary union, while formerly the process of healing took months to complete. Before the use of antiseptics, Volkmann lost 12 out of 16 compound fractures. He has since published, in the 'Proceedings of the Sixth Congress of German Surgeons,' a list of the cases treated antiseptically in the Klinik at Hallé during three years. Amongst these, but extending over a period of four and a half years, are 75 cases of compound fracture treated conservatively, including 1 of the thigh, 6 of the knee, 42 of the leg, 6 of the arm, 5 of the elbow, and 17 of the forearm; all recovered *quoad vitam*. Eight required secondary amputation, from which they too recovered: 4 in the thigh, 1 in the leg, 2 in the arm, 1 in the forearm.

In the whole period of three years, there were 183 amputations or disarticulations in 172 persons, with 23 deaths; but on analysing these cases, we find that in 139 uncomplicated cases of amputation or disarticulation there were but 4 deaths: 2 of these being in four hours after operation, and

1 in twenty-four hours. There were 42 amputations of the thigh, with but 1 death, and that with severe injury of the buttock; it took place within 24 hours. There were 25 amputations of the leg, with 1 death from 'habitual' erysipelas. There were, besides, cases classed as complicated; viz., 9 cases of double amputation, with 2 deaths; both thighs being amputated in each of the fatal cases. Fifteen cases were admitted in a septic condition, with 8 deaths. In 3 other fatal cases the wound was aseptic; 1 of these died from delirium tremens, 1 from pneumonia, and 1 from puerperal pyæmia. 50 osteotomies, besides 9 resections for false joint, took place, with but 1 death, occurring from hæmophilia within 24 hours. Among the osteotomies were 10 cuneiform resections of the femur, and 19 of the tibia.

I understand Professor Volkmann is soon about to publish another report; and Professor Esmarch says his triennial statement will show even better results than those I have quoted.

Dr. Schede's[1] experience in the treatment of compound fractures of the limbs is similar to the results obtained by Volkmann.

Among 37 cases of recent compound fractures treated antiseptically only 1 died, and shortly after the injury from delirium tremens and fat embolism of the lungs.

Amputation was subsequently performed in 4 on account of gangrene. In 3 of them the amount of injury to the soft parts had been under-estimated; the fourth patient was so delirious that he tore through the posterior tibial artery in tossing the injured limb about, the anterior having been lacerated previously. Most of the cases were severe and due to direct violence.

Among the fractures in Volkmann's *résumé*,[2] 21 were

[1] '*Handbuch der Algemeinen und Speciellen Chirurgie*,' von Pitha und Billroth, 1880.

[2] '*Verhandlungen der Deutschen Gesellschaft für Chirurgie*,' 1877.

diagnosed with certainty to involve the adjacent joints, 3 of these required secondary amputation and 7 were excised; while no fewer than 11 recovered, with mobility in the articulation in 10 instances, and anchylosis in 1 only.

Amongst the compound fractures published by Schede were 15 which opened into joints, 1 of them died from tetanus after a laceration involving the wrist-joint, the only fatal case. Two required amputation for gangrene; 1 was excised; the remainder, consisting of 5 elbow-joints, 2 wrist- and 2 ankle-joints recovered with free motion. A tenth case, with most extensive '*colossal*' damage to lower part of leg involving the exfoliation of 40 sequestra, half of which came from the joint surface of the tibia, recovered with anchylosis, but without suppuration in ankle-joint.

Taken together, and excluding the case of tetanus, these cases comprise 35 compound fractures of the joint ends of bones.

19 recovered, either with perfect or almost perfect movement.

2 anchylosis took place.

8 excision was performed.

6 were amputated.

But no single patient lost his life.

These numbers are not large, but under any other form of treatment it may, I think, be safely affirmed that such a result has not hitherto been obtained.

The antiseptic treatment therefore must involve a change in some of the indications hitherto held as demanding primary amputation or excision in compound fracture involving a joint. The need for amputation will in future depend on the extent of the damage to the soft parts rather than on the damage, however great, to the bone.

Primary amputation must only be performed when the injury is so extensive as to leave no hope of saving the limb.

Further, in place of amputation in doubtful cases, we are justified in giving the patient a chance, as we know that the risk is not thus increased, and that if gangrene supervene an intermediary operation can be safely performed, as safely at least as the primary, which is certainly not the case after other kinds of wound-treatment.

Dr. Maclaren, surgeon to the Cumberland Infirmary, has alluded in a recently delivered address [1] to the results of operations lately performed in that hospital, in which he adverts to the fact that the mortality after amputations twenty years ago was not less than 25 per cent., the cause of death in 42 per cent. of the fatal cases being ascribed to pyæmia. During the years 1877 and 1878 amongst 49 amputations of all kinds treated antiseptically there were 4 deaths, 3 of these within twenty-four hours from shock and hæmorrhage, the other from tetanus.

Dr. Bardenheuer [2] mentions in the very interesting account of his practice that the report already published in detail of the previous year 1878 presents a series of results identical with those obtained in 1879, differing only in the fact that during the last year the number of operations has been greater. He has had the astonishing result of not a single death due to the operation amongst 123 cases. These consisted of:

41 amputations, including 17 of thigh.
10 disarticulations, one being at the hip-joint.
53 resections, including 15 of hip and 12 of knee.
23 wedge-shaped excisions of bone.
5 operations for pseudarthrosis.
1 trephining.

He further states that he now finds erysipelas seldom

---

[1] 'Advances in Surgery during the past twenty years, by R. Maclaren, M.D. Surgeon to the Cumberland Infirmary,' *Lancet*, February 1879.
[2] '*Zur Frage der Drainirung der Peritonealhöhle von Dr. Bardenheuer, Oberarzt des Burger Hospitals zu Cöln*,' 1880.

attacks wounds or sores which do not admit of being treated antiseptically, and their better recovery is promoted in a remarkable degree.

Professor Socin of Basel [1] has performed the radical cure of hernia in 17 cases. All were done with strict antiseptic precautions, and proved successful: 10 were inguinal, 7 crural. The neck of the hernial sac was isolated and divided, after tying it with catgut as high up as possible. The pillars of the ring were then usually sewed together, the sac being afterwards removed. In some cases, the excision of a circular strip from the neck of the sac proves sufficient. In no case was there any serious symptom, and so far not the slightest reappearance of the hernia, although, in some, a year or sixteen months had elapsed. In 3 of the inguinal operations, there was previous strangulation: all the femoral herniæ were strangulated. I have myself performed a similar operation. The result was quite successful; but whether the cure be really radical a longer interval of time will show. It must be a matter of doubt how far any form of operation affords a good prospect of definitive cure. In every case of strangulated hernia, in cases of irreducible hernia, or those which cannot be kept up with a truss, or in young persons, such an operation as this would seem to be indicated. The operation is likely to be more successful in femoral than in inguinal hernia, for in this the presence of the spermatic cord prevents the complete closure of the channel.

Dr. Schede, surgeon to the Städtisches Krankenhaus, in Berlin, has also published in Pitha and Billroth's Handbook of Surgery, a series of statistics, comparing the results of amputation, as practised with antiseptic precautions, and without.

The prognosis of an amputation wound is, I hope, changed from the days when Malgaigne showed that the mortality of all amputations, in the Paris hospitals between 1836 and

[1] Langenbeck's '*Archiv für Klinische Chirurgie*,' vol. xxiv.

1841 excluding finger amputations, was 50 per cent. of the cases 601, viz. with 308 deaths.[1]

Paul[2] has collected 11,927 cases of amputation and disarticulation of the larger limbs, with 4,185 deaths, or 35·008 per cent. mortality, that is of every 3 persons operated upon 2 recovered and 1 died. Of course, such collections are very imperfect ones to form any exact conclusions from, as the age, condition of the patient, reason for operation, the method of operation, the after-treatment, the position of the wound made as regards the trunk, its size, and much more besides, being all important elements of success or failure, are not stated. But making allowance for all this, the fact remains that 1 person died in every 3 operated upon.

For injury, the mortality has hitherto shown itself about double that of operations performed for disease.

In traumatic cases the period at which amputation is performed, whether primary, intermediate, or secondary, has hitherto exercised a great influence on the rate of death.

In civil practice it is uncertain whether primary or secondary amputation for injury is the more fatal; but in war practice, there can be no doubt that primary amputation affords, in every position in the body, better chances for the patient's recovery than secondary.

To ascertain the influence of after-treatment on cases of amputation, cases otherwise similar in character must be compared in which the danger to the patient depends on the progress of the operation-wound alone. Dr. Schede has published a valuable series of cases of amputation, collected from the published Reports of German Hospitals during the last 20 years. He contrasts the ordinary methods of treatment, as practised by such men as Von Bruns, Bardeleben,

[1] From a recent *Thèse*, 'Etude Comparative entre les principales Méthodes de Pansement de Plaies,' par M. Briquet, it would appear that Paris mortality with ordinary surgical dressings continues to be as large as before.

[2] '*Die Conservative Chirurgie der Glieder*,' 2nd edit.

and Billroth, with strictly antiseptic cases of his own, and Professors Busch, Socin, and Volkmann.

These cases are arranged in nearly similar groups; the reports from which they are gleaned giving sufficient details to classify them with accuracy.

321 cases of amputation for injury, for disease, and for new growths of the different limbs, treated antiseptically, excluding the smaller amputations of fingers and toes, were followed in 14 cases by death, and of these 14 fatal cases, 8 were due to collapse on the day of the operation, and were

ANTISEPTIC TREATMENT ACCORDING TO LISTER.

	Cases	Recovery	Death	
Amputation at Shoulder-joint.				
For Injury	4	3	1	Death in four hours
For New Growth	5	5	0	
Total	9	8	1	
Amputation of Arm.				
For Injury	20	20	0	
For Disease	12	12	0	
Total	32	32	0	
Amputation of Forearm.				
For Injury	34	34	0	
For Disease	13	13	0	
Total	47	47	0	
Disarticulation at Wrist.				
Injury	3	3	0	
Disease	1	1	0	
Total	4	4	0	

therefore unaffected by any form of treatment. Only 6 fatal cases remain, and one of these was dependent on senile debility, and one on tetanus. Only 3 were attributable to septic causes.

Of a nearly similar number, 387, treated in the ordinary way, 91 or 23 per cent. died from septic poisoning—a cause which we may hope hereafter to call preventable.

As such a contrast has not before been made, I have published the Tables almost *in extenso*.

FORMER METHODS OF TREATMENT.

	Cases	Recovery	Death	Causes of Death
Amputation at Shoulder-joint.				
For Injury	9	4	5	2 Pyæmia 2 Septicæmia 1 Erysipelas
For New Growth	6	3	3	2 Pyæmia 1 Secondary Hæmorrhage and Septicæmia
Total	15	7	8	
Amputation of Arm.				
For Injury	22	16	6	4 Pyæmia 2 Septicæmia
For Disease	19	19	0	
Total	41	35	6	
Amputation of Forearm.				
Injury	20	19	1	Pyæmia
Disease	22	20	2	Death from Marasmus. An old man of 73. Carcinoma of hand.
Total	42	39	3	
Disarticulation at Wrist.				
Injury	9	9	0	
Disease	1	1	0	
Total	10	10	0	

ANTISEPTIC TREATMENT ACCORDING TO LISTER—*continued.*

	Cases	Recovery	Death	Causes of Death
Disarticulation at the Hip-joint.				
Injury	2	0	2	Both on day of operation
Disease	4	2	2	Each four hours after operation
Total	6	2	4	
Amputation of Thigh.				
Injury	23	18	5	{ 1 Septicæmia { 1 Collapse in twelve hours
Disease	63	62	1	Hæmorrhage
Total	86	80	6	
Disarticulation at the Knee-joint.				
Disease	3	3	0	
Amputation of Leg.				
Injury	19	19	0	
Disease	50	49	1	Habitual erysipelas
Total	69	68	1	
Partial Amputations of Foot.				
By the methods of Syme, Pirogoff, Chopart, Lisfranc and Amputation of Metatarsus	65	63	2	{ 1 Old woman 77 twenty-three days after operation. No fever. Wound almost healed. Weakness
8 of these operations were performed for injuries, the rest for disease.				

SCHEDE'S TABLES.

Former Methods of Treatment—*continued*.

	Cases	Recovery	Death	Causes of Death
Disarticulation at the Hip-joint.				
Injury	0	0	0	
Disease	3	1	2	1 Pyæmia 1 Exhaustion
Total	3	1	2	
Amputation of Thigh.				
Injury	24	10	14	6 Pyæmia 5 Septicæmia 2 Collapse shortly after operation 1 Exhaustion
Disease	81	52	29	20 Pyæmia 3 Septicæmia 1 Secondary Hæmorrhage 1 Marasmus 1 Weakness 1 Unknown
Total	105	62	43	
Disarticulation at the Knee-joint.				
Disease	7	6	1	Pyæmia
Amputation of Leg.				
Injury	28	16	12	10 Pyæmia 2 Septicæmia
Disease	87	62	25	19 Pyæmia 5 Septic 1 Hæmorrhage
Total	115	78	37	
Partial Amputations of Foot.				
Injury 2 Cases Disease 37	39	29	10	8 Pyæmia 1 Erysipelas 1 Collapse

Thus in 321 cases of amputation, treated strictly antiseptically, there were 14 deaths, or 4·36 per cent. If 8 deaths on the day of operation be subtracted, we have only 6 deaths, or 1·9 per cent. in this long series of important operations. The following tables show the causes of death :—

Pyæmia	0
Septicæmia	1
Erysipelas	1
Trismus	1
Pyæmia Simplex	1
Secondary Hæmorrhage	1
Senile Weakness	1
Collapse on day of Operation	8 = 14

In 387 cases treated on the old system we have a total number of deaths 110—28·42.

Pyæmia	72	or 18·60	per cent.
Septicæmia	19	„ 4·83	„
Erysipelas	2	„ 0·51	„
Pyæmia Simplex	6	„ 1·55	„
Secondary Hæmorrhage	3	„ 0·77	„
Senile Debility	2	„ 0·51	„
Collapse on day of Operation	6	„ 1·55	„

Anything more strikingly favourable it would be difficult to put forward on behalf of the antiseptic treatment, both as compared with other kinds of treatment in the hands of able and distinguished men, and also as an absolutely good result in itself.

It is very noteworthy that, if the cases of septicæmia and pyæmia be subtracted from this table of deaths, there remain 19 fatal cases, or 5 per cent. of those operated upon—a proportion almost identical with the results, 4·36 per cent., obtained by the antiseptic method, which claims to eliminate septic causes of death.

To these remarkable tables follow others, comparing the results obtained in complicated cases, and in those where

MORTALITY AFTER AMPUTATION.

septic changes had already occurred, also in cases suffering from various forms of intercurrent disease. Here the comparison is also, but not to so great an extent, to the advantage of the antiseptic method.

In the antiseptic cases the gross mortality is 4·36 in other cases of a nearly like number, 28·42. In this first series, there is one fatal case of septicæmia; in the other 91 fatal cases of pyæmia and septicæmia are recorded. The mortality of the cases under the old system as recorded by Schede is not excessive, being 178 in 471 cases of amputation, or 37·8 per cent. from all causes. Paul's average is 35·08.

This is higher than our best English results, of which Professor Spence's may be taken as a fair example, reported by Mr. Cheyne [1] from Professor Spence's papers in the *Medical Times and Gazette* and *Edinburgh Journal*. Of 331 major operations during 5 years, 58, or 17·1 per cent. died—while of 97 major amputations performed during the same period 25 died, or 25·7 per cent.

These consisted of amputations both for disease and injury, as follows:—

	Primary		Secondary to Injury		For Disease		Total	
	No. of Cases	Deaths	No. of Cases	Deaths	No. of Cases	Deaths	No. of Cases	Deaths
Hip-joint	0	0	0	0	3	1	3	1
Thigh	5	3	1	0	25	6	31	9
Knee-joint	1	0	0	0	0	0	1	0
Leg	4	0	1	1	8	1	13	2
Ankle	1	0	1	1	23	1	25	2
Partial of foot.	0	0	0	0	6	2	6	2
Shoulder-joint.	1	1	1	0	2	1	4	2
Arm	2	1	2	0	5	2	9	3
Forearm.	2	2	0	0	2	2	4	4
Partial of hand	0	0	0	0	1	0	1	0
	16	7 or 43·7 per ct.	6	2 or 33·3 per ct.	75	16 or 21·3 per ct.	97	25 or 25·7 per ct.

[1] *British Medical Journal*, February 14, 1880.

The number is not large, but I think a very similar kind of experience would be found to be common amongst hospital surgeons; and in Mr. Spence's 2nd Edit. of his Lectures on Surgery he reports a similar mortality in his previous major amputations, including those up to 1875, 503 in number:

    134 primary with 57 deaths = 42·5 mortality
    37 secondary injury 15 ,, = 40·5 ,,
    333 for disease   68 ,, = 20·4 ,,

A considerable number of the deaths were from pyæmia.

No more remarkable results have been published than those that have followed the treatment of gunshot wounds. Kraske[1] gives the results of 23 gunshot wounds, treated in Hallé. Details of all these cases are given in full, and can be studied in the original paper. Five, of the soft parts, speedily healed. Two others, also, which involved the ligature of the femoral artery; and 2 cases of gunshot fracture of the humerus, recovered. Four gunshot fractures of the knee recovered, 3 having a movable joint—1 a most remarkable case. Of 3 perforating gunshot fractures of the skull, 2 recovered and 1 died. 6 penetrating gunshot wounds of the thorax all recovered. An abdominal gunshot wound, injuring the liver, kidney, and pleura, proved fatal. These results were achieved by primary disinfection of the wound, with enlargement of one or both openings, sufficient drainage, and complete antisepsis; in the thorax cases by antiseptic occlusion.

The case alluded to as remarkable, was presented by Volkmann to the Sixth Congress of German Surgeons, at its meeting in 1877, two months and a half after the receipt of the injury.

The subject of the accident was a surgeon, twenty-four years of age. He was shot during a duel in the upper part of the right tibia. The bullet entered on the outer side of

[1] Langenbeck's *Archiv*, vol. xxiv., Part 2.

the limb, and lodged in the tibia, after having traversed almost the entire width of its expanded upper part. The sound passed for 5·8 Cm., upwards of two inches, into the bone before it touched the bullet.

Synovia flowed copiously from the wound, and the joint became tense from effused blood. There could be no doubt that it was involved in the injury.

The removal of the ball was very difficult. A large funnel-shaped channel had to be made in the head of the tibia, as indicated in the accompanying diagram, to give room for the instruments used in extracting the ball. A

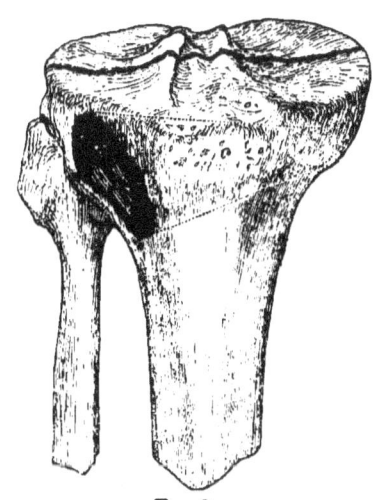

FIG. 3.

fissure opening into the joint cavity could be seen in the roof of this channel. Through this air entered the joint and caused an audible sound during the use of the chisel.

On removal of the ball the depth of the wound in the bone was 6·4 Cm. In order to clear the blood out of the joint an incision was made into it two Cm. long at the internal border of the patella, the joint was washed out, and a drainage tube inserted, as also one into the cavity in the bone. Lister's dressings and a splint were applied.

There was never the slightest local reaction. For the first four days a slight elevation of temperature took place, but afterwards there was no fever. On the fourth day the drain was removed from the joint, because no secretion flowed from it even on pressure. On the sixth day, after five times twenty-four hours, the drain was removed from the wound in the bone. This was filled with a dark firm coagulum, which underwent the usual changes. The passage left by the drain was found closed at the next dressing. There was never a trace of suppuration. A month after the infliction of the injury, the superficial layer of the coagulum was cast off and the large bone cavity found filled with organised tissue, partially cicatrised on the surface. In forty-six days the patient left his bed, two days later he could somewhat bend the joint and get about with a stick. And at the time he was shown to the members of the Surgical Congress, only two and a half months after his severe injury, he was able to bend his knee to a right angle, climb the steep steps of the theatre in which the meeting took place, and walk about all day long.

Professor Estlander has published in the 'Nordiskt Mediciniskt Archiv' for 1879, accounts of a series of head injuries treated in the Hospital at Helsingfors antiseptically, as compared with a number previously treated by ordinary methods. There were treated in the ordinary way.

From 1860 to 1869:

	Cases	Deaths.
Simple scalp wounds	82	3
Scalp wound, with exposure of Skull .	37	7
Compound-fracture of Skull . .	10	7
Fracture of Base of Skull . .	8	4

From 1870 to 1877:

Antiseptic Treatment.

Simple scalp wounds	95	3
Scalp wounds with exposure of Skull . .	67	1
Compound-fracture, with injury of Brain .	13	2
Fracture of Base	11	5

A comparison of these two tables shows that in the severer forms of head injury involving the bone, there was a great diminution in the mortality of the cases treated antiseptically. In the cases of scalp wound, with exposure of the skull, and in cases of compound fracture with brain injury, the difference is striking. In the former the death rate is as 19 per cent. compared to 1·5 per cent., while in compound fracture of the skull, three-fourths of the patients died in the 1st period, and one-sixth in the 2nd. In fractures of the base there is no difference, nor is there any in simple scalp wounds.

Some remarkable observations, made during the late Russo-Turkish war, in the army of the Caucasus, are published by Dr. Carl Reyher,[1] Consulting Surgeon to that division of the Russian army.

He found it absolutely necessary, in order to obtain perfect antiseptic results, to place the wounded under treatment immediately after the injury; and that it mattered but little, as regards the final issue, whether the antiseptic treatment was adopted at a subsequent period, or the wound not treated antiseptically at all. From an elaborate number of tables which he publishes, I can only make a few selections. The following two are illustrative cases.

A man shot through the internal condyle of the femur, on October 3, on the heights of Awlijar, sustained a fracture with effusion of blood into the knee-joint. The ball was lodged in the condyle. Complete recovery with a movable joint took place in six weeks, under antiseptic occlusion from the first. Five dressings were made. The highest temperature was 100·4. The patient was thrice transported distances of 15, 40, and 210 kilomètres.

A second case was that of a man wounded before Kars. The patella was comminuted, and the ball could be felt

[1] Volkmann's *Sammlung Klinischer Vorträge*, August, 1878.

lodged in the tibial joint-surface by the finger introduced into the joint. A much deformed Martini-Henry ball was extracted; the joint was washed out with five-per-cent. carbolic acid solution, and a large drain introduced. The wound remained aseptic throughout. Only once there occurred a slight rise of temperature, otherwise the temperature remained normal. Complete recovery took place with a movable joint. This patient was also transported long distances—20, 40, and 210 kilomètres. Of wounds of the larger articulations, the shoulder, elbow, wrist, hip, knee, ankle, and tarsal joints, there were 186 cases.

Penetrating Gunshot Wounds of the larger joints: Shoulder, Elbow, Wrist, Hip, Knee, Ankle, and Tarsus.

	No. of Cases	Deaths	Mortality per cent.
TABLE X.—Cases treated antiseptically from the outset, either by occlusion, antiseptic drainage, or primary resection	46	6	13·0
TABLE XI.—Cases treated by secondary antisepsis	78	48[1]	61·5
TABLE VI.—Cases treated without antiseptic precautions. Conservatively	62	39	62·9

During a second visit to the Caucasus, Prof. Reyher found that amongst the last category of 62 cases nine more had died, making forty-eight deaths, or a mortality percentage of 77·4.

TABLE XII.—*Eighty-one Gunshot Wounds of the Knee-joint, treated conservatively.*

	Number of Cases	Deaths	Mortality per cent.	Number of times the Extremity was Preserved	Number of times Mobility was Preserved
Primary Antisepsis	18	3	16·6	15	15
Secondary Antisepsis	40	34	85·0	1	0
Non-Antiseptic Cases	23	18	18·2	2	1

[1] Twenty-nine from Septicæmia and Pyæmia,

As a final result of the 23 Non-Antiseptic cases in this series only one individual has survived, which raises the mortality almost to its highest point.

Contrast in Fractures treated Conservatively.

	Cases	Deaths	Mortality
Gunshot Fractures treated by primary antisepsis	22	4	18·1
Gunshot Fractures otherwise treated	65	23	35·3

The following table shows the contrast in respect of frequency of Septic Diseases in the two classes of cases.

	Cases	Deaths from Pyæmia and Septic Phlegmon	Mortality
TABLE IX.—Gunshot injuries to Bone, including Fractures, injuries of Joints, and amputations, treated by Primary Antisepsis	81	5	6·1
Similar injuries treated by Secondary Antisepsis	143	46	32.1 Five times as many

A similar proportionate amount of pyæmia and septic abscess occurred in wounds of the soft parts treated in the two ways.

The results obtained by Reyher in his treatment of gunshot injury of the knee (Table XII.) are certainly the most remarkable on record, and deserve the attention of every military surgeon.

Compare these with what obtained in our own army in the last great war in which it was engaged. I speak with the authority of Surgeon-General Longmore when I state that in no single instance during the Crimean war was a knee-joint which had been opened by a bullet saved, life being lost in every case where amputation was not resorted to; while in Professor Reyher's cases the limb was not only saved, but the functions of the joint preserved as well.

Of 18 primary antiseptic cases of gunshot wound of the knee, in 15 cases the limb was preserved with a movable joint, and none required amputation; whereas 60 per cent. of the cases of injury of this joint treated in the ordinary way, or by secondary antiseptics, required amputation, and 84·5 per cent. of the persons amputated, died.

	Primary Antiseptic.	Secondary or Non-antiseptic.
Cases	18	63
Deaths	3	52
Mortality per cent.	16·6	83·1
Limb preserved in	15	3
Mobility preserved in	15	1
Without movement in joint	0	2

Such statements as these may be disbelieved by some, but until they are contradicted it would surely appear to be our plain duty to consider the importance of their bearing upon the treatment of wounds.

Antiseptic dressings have been employed in burns with the effect of limiting the destruction of tissue to that actually involved in the injury, diminishing the pain and fever, reducing the amount of suppuration to a minimum, the cicatrix afterwards formed being unusually smooth and stretchable.

I cannot now allude to the more extended employment of laparotomy in abdominal diseases in the hands of Czerny, Billroth, and others, nor to the admirable results of Professor Rossander of Stockholm, in cataract operations under carbolic spray.

It is open to us, I think, to believe, without being accused of prejudice, that Lister's method has shown itself to be, up to the present time, the best available antiseptic method; that it prevents more perfectly than any other method that chief danger of wounds—putrefaction, and the diseases which arise from it.

In a brilliant and eloquent address, to which I had the privilege and great pleasure of listening, Mr. Savory [1] denies that Lister's method is better than other methods, asserts that it delays the healing of wounds, and that it fails in a certain number of cases. From my own experience, I believe it to be better; and I have made an effort to show that the results are superior, but it is no easy task to traverse the positions of the persuasive orator of the last meeting of our Association. It is true the method from time to time has failed; but it is fair matter for argument whether the failure be due to the weakness of the method or the faulty manner of its application. The protracted employment of carbolic acid will interfere with the cicatrisation of a granulating surface; but when the surface is granulating is just the period in the history of a wound when, as a rule, a strict antiseptic occlusion may be safely dispensed with. That operations have been done, and successfully done, with the aid of Lister's method which, in Mr. Savory's opinion, had better have been left undone, is a high compliment paid to the method at the expense of the surgical discretion of the operator. The wound is shut up from view, no doubt, but in a manner which secures it complete and desirable rest, while it may be inspected with ease and safety at any time, if for any reason this be required.

Those who adopt this method do so, I suppose, because they find it produces for them the best results. They have been accused of losing sight of the principles of wound-treatment in the pursuit of frivolous details; whereas the carrying out of this method is pervaded by the presence of a great principle, the prevention of wound-putrefaction and its consequences, and this it is which animates details which

[1] Address in Surgery, British Medical Association meeting at Cork, 1879.

would be otherwise repulsive and meaningless. I should like to put these questions. What is the cause of wound-putrefaction and wound-diseases? Are the results of Lister's method obtainable by other methods? or, in other words, Is Lister's method superior to other methods?

It is as a contribution, however imperfect, towards the solution of these propositions that I have prepared this Address. The strict antiseptic method has in effect been characterised by some as 'not new,' and by others as 'not true.' It is time that some wider and more definitive expression of opinion on its merits shall be made.

In conclusion, I wish to say that I recognise in the fullest manner that the care and dressing of a wound is but one part of a surgeon's duty; that he has many other important functions, to neglect which is to invite failure. In this paper I have only considered what are the means best calculated to obtain that surgical cleanliness which is almost the equivalent of surgical safety; and I have tried to show cause for believing that this is only to be achieved when the wound—be it large or small—is maintained throughout free from decomposition, and especially is so maintained during the earlier periods of its existence, when it is suffering from the depressing effects of the recently inflicted injury; its surface prone to decompose from the violence inflicted on it, and the vital reaction which produces a protecting shield of living cells, able to a certain extent to resist noxious influences from without, still incomplete.

Surgical safety, and all the great consequences which follow from it, is the claim which is put forth for Lister's method of dressing wounds. Is there any other method, even that with the 'simplest, safest, best of antiseptics,' clean water, which will permit the surgeon to view a mass of dead blood lying in an open wound being transformed day by day before his eyes into living tissue? I do not mean the blood is vivified, but amœboid cells invade it, eat it up, as it were, and flourish in its stead. If this be true, and I suppose its truth will not be denied; if the great joints may be opened with absolute impunity by the surgeon's knife; if such formerly fatal injuries as gunshot-wounds of the knee can be saved amidst the difficulties of an army in the field; if ovariotomy be made more successful in the hands of experts, who operate on such cases by hundreds, and what is, to my thinking, more important, the operation is rendered safer and more successful in the hands of all surgeons; if, in injuries to the head, in ordinary amputations and excisions, and operations of various kinds, and in compound fractures, the risk to life be not only diminished, but the recovery of the patient marked by a minimum of fever, pain, and suppuration, then a gain has been accomplished for surgery which it is scarcely possible to characterise; and I do not know which to admire the more—the scientific mind which has grasped a great principle and applied it, or the character of the man who has unswervingly pursued the object of his life, patiently perfecting one by one the means adapted to procure the end in view. To John Hunter we owe our idea of the harmlessness of subcutaneous injuries and all the great advantages to surgery that are derived from the application of this knowledge; and if in time our ordinary surgery shall become as safe as subcutaneous surgery, this we shall owe to Lister.

DEBATE.

MR. BRYANT: I premise that we have met this evening to consider the results of antiseptic surgery rather than the theory on which the practice was based, and I warmly thank Mr. Lister for his untiring and able advocacy of the antiseptic treatment, as well as for having shown surgeons the necessity of attending to many points of practice which previously they were disposed, perhaps, to regard as trivial and unimportant. Indeed, we must feel that it was to him, directly or indirectly, in a very great measure, that much of the recent improvement in surgery was to be attributed.

In saying this, however, it was not to be forgotten that great improvement in hospitals and in surgery had been gradually going on during the last twenty-four years, and that these results had been brought about by many combined causes—such as a better understanding of hospital management and more particularly of hospital hygiene; a clearer comprehension of the conditions and causes of disease; a keener appreciation of the value of physiological and mechanical rest, as enforced by Hilton; a fuller recognition of the importance of efficient drainage and of complete cleanliness of wounds and the introduction into practice of the torsion of arteries with the consequent discarding of silk ligatures.

The introduction of antiseptics formed a crowning triumph of surgical improvements, and for this we are unquestionably indebted mainly to Mr. Lister.

The fact, however, must be recognised that the improvement in surgical practice which has been mentioned was found and seen in hospitals and in wards of hospitals where 'Listerism' was *not* practised, as well as in others in which it was systematically followed—and I would here ask what is antiseptic surgery? and if it is to be admitted that there is any form of surgery or of dressing wounds to which this term can be rightly applied outside that followed by Lister? For my part I think there is, and that any system of treatment which has for its object the prevention and neutralisation of putrefactive changes in wounds might fairly be called antiseptic. Indeed, with such a view I claim to be an antiseptic surgeon, although I may not always use the spray; and from what has recently been written upon the subject I am led to think that this view will soon become more general, and that the followers of Lister are ready to adopt it.

I will briefly consider some of the points claimed by Listerism as being peculiar to the system, such as the absence of fever after operations; freedom from pain and inflammation and suppuration of wounds, and ask if these advantages really belong wholly and solely to the system? I answer emphatically that they do not; and to prove my point state that out of a series of 50 operations of my own including severe capital operations of all kinds, in only 3 was there any elevation of temperature beyond the third day, whilst in all the others the temperature, after the first rise which followed the operation, fell to about normal, where it remained during the period of convalescence,—and this experience I do not believe to be in any way singular.

I will now pass on to criticise a paper recently published by Mr. W. Watson Cheyne ('British Med. Journal,' Nov. 29, 1879), including *all* Mr. Lister's operations on healthy joints performed under the spray (a period of eight years), together with such accidental wounds of joints as occurred during the same period, the cases, as the reporter describes, being a *selected series* of cases which must be looked on as *crucial tests* of the treatment; and I must express the surprise and disappointment I felt when on reading it I found that the oft-repeated assertion that 'loose cartilages may *now* be removed by direct incision from the bone without any apprehension of evil results,' was based upon the experience of only *three* cases; and another assertion no less startling, 'that a joint may be cut into with the *certainty* that no danger will follow, *with perfect safety* and without risk,' was based upon the limited experience of another 16 cases. This surprise was also much enhanced when I found that several of the cases were of a trivial nature, some questionable and others satisfactory; the whole when read without a bias possibly suggesting that the practice which had been adopted in their treatment was good, but no more, since they certainly by no means afforded conclusive evidence as to its advantages. Indeed, I am disposed to think that an equally good series of cases might be extracted from the ordinary case-book of the hospital surgeon.

In saying this, however, I by no means wish to assert that evidence is wanting as to the value of antiseptic surgery as practised by Lister—for the reports of foreign as well as of home hospitals afford ample testimony as to its great value, and more particularly where cleanliness was not studied and hospital hygiene was unknown; but I condemn the strong assertions that have been made upon limited data, and the positive generalisations which have been drawn from a narrow experience. The theory of the practice may be sound and the practice itself successful, but neither the theory nor practice will gain a point by strong dogmatic assertion unsupported by equally strong facts.

MR. C. MACNAMARA: It seems to me in dealing with the subject of antiseptic surgery, or as I should more correctly say of Listerism, we must recognise the fact that septicæmia, and pyæmia differ from one another, as much as typhus, does from typhoid fever. Septicæmia depends on changes in the blood which sometimes occur independently of the atmosphere; for instance, a person may receive a blow on the head, with no external wound, but in consequence of the injury a portion of the dura-mater is separated from the skull, pus forms between this membrane and the bone. In a case of this kind I have seen rapid and fatal septicæmia follow, the heart's action becoming extremely feeble, and the respiratory muscles almost paralysed, the patient dying within twelve hours of the commencement of these symptoms. After death clots were found filling the right side of the heart; plugging of the capillary vessels, with spots of effused blood round them were scattered beneath the various serous and mucous membranes of the body; these plugs are formed of organic matter, probably derived from dead white blood-corpuscules, destroyed by means of the influence, (whatever it may be) which in the living body causes septicæmia. Multiple abscesses are never found after death from uncomplicated septicæmia, the patient has no marked rigors as in pyæmia, but suffers from continued fever in proportion to the severity of the case; a large number of the patients so affected recover—for septicæmia varies in intensity from instances such as those I have referred to, to cases of ordinary surgical fever.

Pyæmia, so far as our surgical cases are concerned, depends upon impure air passing into the open vessels of a wound and the connective-tissue spaces, and exciting putrefaction in the organic matter they contain; fragments of this decomposing material are carried into the circulation, and being arrested by the smaller vessels in various organs, it excites in these organs, similar action to that which the putrefying matter is undergoing, thus producing what are called necrotic abscesses. From these abscesses more decomposing matter is carried to other parts of the body, and so the patient's blood becomes poisoned, a succession of abscesses occur, and it is extremely rare for a case of this kind to recover. Pyæmia is characterised by intermittent fever, rigors being a prominent symptom; at the commencement of the disease the temperature of the body rising and falling in a very remarkable manner. It frequently happens, however, that septicæmia and pyæmia run their course simultaneously in the same individual to a fatal termination (septo-pyæmia).

If these premises are true, supposing we could surround a patient during an operation, or while dressing a fresh wound, with an atmosphere containing an antiseptic material, it is evident that the open vessels and connective-tissue spaces of the wound would suck in this antiseptic atmosphere, and its presence would prevent the decomposition of the organic matter within the sides, and the surfaces of the wound; and if you subsequently dress the wound so that no air can gain access to it unless it is saturated with an antiseptic material, the wound will heal without any decomposition taking place in, or about it; and so you would prevent pyæmia from occurring.

We know from experiments undertaken for this purpose, that decomposition will go on more rapidly in a solution of organic matter if exposed to an impure, than to a pure atmosphere; in fact, by filtering the air passing to solutions of organic matter, or keeping the solution in perfectly pure air, you may prevent putrefaction from taking place in them. It is evident, therefore, that if the air of an operating theatre or the wards of a hospital is impure, that pyæmia is particularly liable to occur; because an atmosphere of this description is very likely to set up rapid decomposition in the organic matter of a wound, which is absorbed into the circulation before there has been time for the surrounding living protoplasm to build up a protecting medium about the wound, so as to preserve the system from contamination.

If, therefore, you could always operate in perfectly pure air, you would still meet with occasional cases of septicæmia, for under certain conditions of the system septicæmia will occur independently of the atmosphere, but under these conditions you would not meet with cases of pyæmia; consequently, the greater the care taken to render the atmosphere surrounding your patient free from organic matter, the less the necessity for antiseptic surgery. In by far the larger part of the world, however, it is impossible to secure a pure atmosphere. It is simply foolish for medical men to talk about pyæmia being a disgrace to the surgeon in charge of the case; there are many conditions under which we must combat with disease very different from those regarding which practitioners in this favoured land have to contend.

I allude particularly to tropical countries, and to a very large proportion of hospitals built in Continental cities. But it is evident from the remarks I have already made, that the worse the hygienic conditions under which our patients are placed, the greater the triumphs of antiseptic surgery should be, for this system is intended to correct the impure condition of the atmosphere by which they are surrounded. It seems absurd, therefore, to argue, because wounds do well in a few hospitals, in which the patients have the advantage of the best possible surgical aid, admirable nurses, and where expense is a matter of no consideration, that Mr. Lister's system is not invaluable to patients placed under less favourable conditions. From my own experience in a large Calcutta hospital, in spite of the most careful dressings, including that which is generally described as 'dry dressing,' it was almost impossible to preserve patients suffering from compound fractures of the lower extremities from pyæmia; and those who have not practised in the tropics can have no conception of the terrible ravages which pyæmia, at certain seasons of the year, effects among surgical patients of all classes. When first Mr. Lister's system of dressing was brought to my notice we employed the hand spray; but I can safely assert that in proportion to the perfection with which his system of dressing wounds was carried out, so was our success. Before leaving India, compound fractures, such as those to which I have referred, were treated one after the other under the Listerian system, with as much freedom from pyæmia as our simple fractures had been. Dr. K. M'Leod informs me that at the present time, Mr. Lister's method of dressing wounds is practised with the very best results in the surgical wards of the Calcutta Medical College Hospital.

If in London, or any other place, you can secure results with which you are satisfied without antiseptic surgery, there seems to be no valid reason why you should resort to this system: but do not allow experience of this kind to influence your judgment, with reference to the enormous advantages which are to be derived from the Listerian system of dressing under less favourable conditions.

Mr. Lister has been more than once challenged to produce statistics, as to the results he has achieved in operative surgery when employing his system of antiseptic dressing, in order that his figures might be compared with those of surgeons who avoid Listerism. I feel sure he must know that it is unwise to accept a challenge of this kind; the results of operative surgery cannot be reduced to figures, for no two cases are exactly alike, the age, the constitution, the habits, the season of the year and many other circumstances exercise their influence upon the condition of patients. With reference to operations, one surgeon may be a more skilful operator than another, or consider himself justified in operating on cases which another man would not think of touching. English surgeons of repute have, as a rule, preferred to allow their practice to be judged by other considerations than those of figures; for instance, there is probably no operation which might be more safely referred to statistics than that of the extraction of cataract, and although we have numerous reports from various foreign surgeons of eminence regarding the number of cases they have successfully operated on for the cure of cataract, we find throughout the long and eminently successful practice of men like Bowman, Streatfeild, and Lawson, that no statistics of their operations have, so far as I know, been published; nor is their reputation, or that of Mr. Lister's, the least impaired by abstaining from statistical reports such as I have referred to.

After the most careful consideration of the subject, I believe Listerism has undoubtedly diminished the mortality and the danger, to patients situated under unfavourable hygienic conditions after surgical operations; and holding this belief, we may argue from the greater to the less, and assert our conviction that this same system is capable of preventing septo-pyæmia in patients under almost all circumstances. I hold, therefore, that Listerism is destined to become more largely employed; for, however difficult the details, or length of time required to carry out these principles in individual cases, it must and will prevail, because it guards our patients from unquestionable dangers.

Mr. BARWELL: I am glad to hear from Mr. Macnamara the effects of antiseptics on the hospitals in the tropics. I, myself, have known of hospitals on the Continent in which after operation death from blood-poisoning was the rule rather than the exception. Now septicæmia and pyæmia are almost unknown in those places. I feel, moreover, that even in what Mr. Macnamara calls wealthy hospitals, provided with every hygienic appliance, this treatment could not be otherwise than beneficial. I will take as examples amputations at the thigh or excision of the knee. Previously to using the antiseptic method I, after such operations, always looked for and very generally found, a certain degree of wound-fever, the thermometer varying from 100 to 102·5. With antiseptics, though I still look for, I do not find such pyrexia, case after case passes with no abnormal temperature, save that which after inhaling ether always attacks a constipated patient on the third day, and which may always be removed by a purge. And what does traumatic fever mean but absorption into the blood of a poison from the wound? it is, in fact, very closely related to septicæmia, only the poison is not quite so strong, not as yet quite but only partially putrescent—it is only undergoing but has not yet undergone the septic fermentation.

I will take another class of operation :—Within the last three years I have five times tied simultaneously the right carotid and subclavian, and in four of the cases with good results. Such success could only be obtained with the aid of antiseptics—the very ligature is part of asepticism. Indeed, following this method further, I have introduced and used a flat organized ligature, which leaves all the coats of the vessel uninjured. Other surgeons have also used it, and in one case (Mr. Holmes) the ligature was applied close to the sac. This ligature, an antiseptic appliance, would I firmly believe obviate the danger, so great, in certain situations, of secondary hæmorrhage from the site of deligation.

Mr. SPENCER WELLS said : For twenty years past I have been most anxious by every possible means to obtain as good results after surgical operations as could be obtained both in hospital practice and private practice, by avoiding or removing all known causes of excessive mortality, carefully excluding every source of poisoning one could possibly exclude, keeping the patient under the very best hygienic conditions that could be obtained by the use of a separate room or the wide division of beds in a ward, and all the other means which sanitary knowledge had placed at our disposal. And I believe that all this care has led gradually to a considerable and general improvement in surgical statistics at home and abroad. Yet improvement did not go beyond a certain point; say 10 per. cent. mortality after ovariotomy. But after the introduction of more complete antiseptic precautions, a very marked and rapid improvement has been observed in results. Perhaps my own experience of ovariotomy may assist in inducing others to believe this statement to be correct, and to follow my example not only in taking the greatest possible ordinary care, but also in gaining for patients all the advantages which antiseptic surgery gives them. Year after year in successive series of 100 cases of ovariotomy, the results had been gradually improving, first from a mortality of 34 per cent., to that of 28 per cent., then 22, afterwards 14, and finally 10 per cent. The mortality had, in fact, been steadily and gradually diminishing year after year with increasing experience and care. But I was not satisfied, and hoping for still better results, I began in 1878 to carry out the Listerian practice in all its details, and in the last 84 cases since adopting these antiseptic precautions, I have found my success increasing in a much greater and more rapid proportion than previously. In the very last series of cases which have been going on under my care, I had 38 cases of ovariotomy running without a single death. During that run there had been 5 cases of hysterectomy. That made 43 cases of large operations for the removal of abdominal tumours without one single death in that long series. Before adopting antiseptic precautions, I did once have a run of 27 cases, and twice runs of 21 cases, and that was the reason why I hesitated and feared that I might not do well to change, and why I delayed longer than I should otherwise have done before adopting a new method. But when I came to compare the results of the antiseptic with those under the former method of treatment, I found they were even more favourable to antiseptics than I stated the other day at the Royal Medical and Chirurgical Society. I have gone over the 84 cases in private practice which I had before commencing antiseptic surgery by the use of carbolic acid spray, and compared them with the 84 cases which followed since, and I have to confess that the results were startling to myself. The statistical result is, that of the 84 cases since adopting carbolic acid spray in ovariotomy (all in private practice), there had been only 6 deaths, and 78 recoveries; whereas of the previous 84 cases in private practice, with all the care I could give to them, there were 21 deaths, leaving only 63 recoveries. And as I went on and became still more accustomed to the method and details of antiseptic treatment, and avoided mistakes, then I obtained the long run of 38 cases without a single death; and, adding to that the 5 more of other important abdominal operations, I can record the gratifying and almost incredible result of 43 cases of these great operations without a death; and these facts I think afford more evidence in favour of Professor Lister's system than volumes of theoretical matter.[1]

MR. THOMAS SMITH: Mr. Lister's method of dressing wounds claims to have abolished certain diseases of a septic character from the practice of operative surgery, and in searching for proofs of the efficiency of the system in this respect, it seems to me reasonable that we should seek out instances of hospitals in which these diseases are rife and there observe the effects of the antiseptic method.

In a well-ordered English hospital there is comparatively little scope for a striking result from the adoption of Listerism, for in these the diseases which Listerism is said to abolish are but rarely observed. For a good test of the value of the system we should go to a hospital where pyæmia, hospital gangrene, and erysipelas are not occasional visitants, but have acquired a right of domicile and never wholly leave the building.

An account of such a hospital is given us by Professor Nussbaum, in his recent publication on Lister's antiseptic dressing. He tells us of his Clinical Hospital at Munich, that pyæmia always flourished there, that since 1872 hospital gangrene has been a constant though unwelcome guest, so that in one year 80 per cent. of all wounds were attacked. He speaks of erysipelas and hospital gastritis as so prevalent in the building that it was the exception for a patient to escape an attack. The hospital was a veritable pest-house.

After trying every kind of dressing without any markedly beneficial effect, on the 1st of January 1875 Lister's treatment was adopted for every wound in the hospital, great and small, and from that day, says Professor Nussbaum, hospital gangrene, pyæmia, and erysipelas were seen no more. As it appears to me, this was a crucial test of the powers of Listerism. Mr. Bryant has criticized severely the statistics recently published in the 'British Medical Journal' by Mr. Cheyne, but though some of the cases are trivial, yet many are instances of such unusual recovery from very severe injuries as to afford striking evidence in favour of the value of Lister's method.

---

[1] The run of 38 cases of successful ovariotomy was afterwards increased to 41, and then a death occurred where septic symptoms had set in before operation.

The statistics presented by Mr. Mac Cormac, for those who demand this kind of proof, should be conclusive as to the superiority of the method in the treatment of gunshot wounds. The groups of cases contrasted occurred in the same campaign, under the same surgeons and in the same class of patients, the injuries being of the same character. For similar reasons Mr. Spencer Wells' statistics afford strong evidence in favour of Listerism.

I venture to call attention to the influence which Mr. Lister has exercised on the practice of those who neither accept his theory nor follow his system, and to point out how much we all are his debtors. For, in addition to a general impulse in the direction of antiseptic surgery, we owe to him certain specific improvements in the treatment of wounds, and chiefly I would refer to the discovery and introduction of carbolized ligatures, whether silk or catgut; to wound-drainage; and to the various forms of carbolic acid now in use.

In the days of silken and hempen ligatures, even ordinary cleanliness was impossible in wound dressings, and the facilities for blood-poisoning were on this account considerable. Wound-drainage was not invented by Mr. Lister, yet he first insisted on its paramount importance, and by his advocacy it was brought into general use.

The same may be said of carbolic acid and its compounds, which, except for his advocacy and influence, might have remained for another generation unknown or rather unused by surgeons. In fact the most strenuous opponents of Listerism avail themselves to a greater or less extent of these improvements to which I have referred, and these things are truly parts, and very large and essential parts, of the antiseptic method.

MR. TIMOTHY HOLMES said: Statistics might or might not be useful, according to the nature of the statistics themselves; but the leading facts of the cases on which the statistics are based must always be given, in order that they should have any value whatever. A simple statement of the gross results—*i.e.* the mere death-rates of lists of cases on either side were in his opinion utterly useless, and worse than useless, for in many cases they were employed to sustain false doctrines. Consequently, I cannot attach any value to the statistics produced by Mr. Mac Cormac from the experience of a foreign army surgeon, since no particulars were given of the cases treated with and without antiseptics.

On the other hand, statistics such as might be gathered from a Hospital like King's College, where two or three surgeons practised two or three different methods under similar circumstances, and upon cases in the long run similar, were, or might be, of very great value if they were accompanied by the leading details of the cases; and I cannot understand why such statistics are not published, or, at any rate, in course of preparation. They would do more, I think, to convince hospital surgeons generally than anything else could do. It was quite true that, since Mr. Lister began to publish and insist upon antiseptic treatment, all surgeons have become to a certain extent antisepticists; and we all recognise the inestimable blessings which Mr. Lister has conferred on the practice of surgery and upon ourselves as practitioners. But when we are told that, in order to practise antiseptic surgery, we must believe in the germ theory, then I cannot but say that belief is not a voluntary act; it must follow upon proof, and no convincing proof of the germ theory as applied to living tissues and living phenomena has, as far as I know, yet been offered. It must be recollected that the debate on which we are now engaged relates not so much to the merits of the antiseptic principle generally, but to the relative merits of Mr. Lister's plans, based on the germ theory, and the common and simpler methods which are now in general use. It has been amply proved that in our London hospitals, under such methods of treatment, surgery is practised with very great success, that wounds heal kindly, that blood-poisoning is very rare, and that in fact there is no reason to think that cases do worse in hospital than in private practice.

I differ, therefore, altogether from Mr. T. Smith in believing that in order to test the merits of Lister's method we ought to go to the foul and neglected wards of some foreign hospital. On the contrary, such a comparison could show nothing bearing on the argument. If the atmosphere of the hospital is loaded with the products of putrefaction, if cleanliness is neglected in the surroundings of the patient, in the materials of the dressings, and very likely in the persons of the dressers, no doubt the antiseptic system either of Lister or of anyone else will effect a great improvement. But this will show nothing as to the relative value of the Listerian and other antiseptic systems. And it ought not to be forgotten that such statements as are now published of the great improvement effected in the results of operations in such neglected hospitals by the antiseptic system have this great disadvantage, that they concentrate attention exclusively on the treatment of wounds and operations, leaving all the other inmates of the hospital to suffer as before from the results of hospital maladministration.

What might happen in tropical climates, as for instance in India, I cannot tell. There might be reasons connected with the climate or the habits of the natives which might prevent hospitals from being made healthy; and there it might, perhaps, be unavoidable to submit to such a state of things as Mr. Smith had described to be characteristic of a Continental hospital; but in France, Germany, and other temperate climates, could there be any reason why a hospital should be allowed to remain in such a state? The surgeon who presided over it seemed calmly to submit to that state of things, and thanked the world that such a man as Lister had been born, in order that he might preserve the minority of his patients, allowing those evil influences to go on, to the detriment of the majority, because he either allowed himself to remain ignorant of the common principles of hospital hygiene, or would not remedy things which, at any rate in this temperate climate, were perfectly remediable.

I hope that nothing in what I have said will be misconstrued, as if I was opposed to the use of antiseptics in surgery, or insensible to the great services of Mr. Lister, who was indisputably the first, if not to invent, at any rate to introduce into general practice, both the antiseptic method and the drainage of wounds. Whether the special developments given by him to the antiseptic methods in obedience to the germ theory should hereafter hold their place as essential to the successful practice of surgery, or should be set aside as superfluous, no lapse of time I believe would ever obscure the merits of the author of the method itself.

Surgeons owe to him a debt of gratitude which they can never repay, which it is a comfort and a pleasure always to acknowledge. The practice will last as long as surgery lasts, and so long shall we thank him.

PROFESSOR LISTER.—I cannot allow the remarks which my friend Mr. Holmes has just made with reference to my friend Professor Nussbaum to pass without a word of remonstrance. I understood Mr. Holmes to say that Professor Nussbaum had been content with the fearful condition in which his hospital was before the introduction of strict antiseptic treatment. Such is very far from being the case. Professor Nussbaum suffered agonies—mental agonies—from the condition of his hospital. He has two such institutions under his charge—one the great Allgemeines Krankenhaus, the other a large private hospital, and the latter was almost free from hospital disease. In the great Allgemeines Krankenhaus he tried, as he states in the work referred to by Mr. Smith, various applications, carbolic acid and other antiseptics. He also tried the open treatment. He used these and other means with the utmost zeal and energy in the desire to get rid of the pestiferous condition of his wards, but he failed. What the explanation of the failure may be it is not necessary for me to inquire into; but what I wish to say is this, that Professor Nussbaum was not indifferent to the conditions under which he worked, and that he was most earnest in his attempts to get rid of those conditions. But then, further, I must protest against regarding with absolute indifference Professor Nussbaum's present results; because he does not only say that he has made an immense improvement on the previous pestilential condition of the hospital, but he tells us he has got a present condition which is superior, I venture to say, to that of any hospital which has not introduced strict antiseptic treatment. He tells us, that since he introduced strict antiseptic treatment he has not had one case of pyæmia, not one case of hospital gangrene (there having been previously 80 per cent.), and that he has abolished erysipelas altogether. Do you believe this statement? In his last edition he indignantly denies the possibility of falsehood on his part in this matter. He has been charged with exaggeration, and points out how such misrepresentation would be not only criminal but impossible; seeing that the Professor of Pathology would not be so civil as to conceal cases of pyæmia if they occurred, this Professor having the duty of performing all post-mortem examinations. Now, Sir, if we are told that these diseases have been banished from the hospital, we must not say that this is a fact of no consequence because that same hospital was previously a pest-house. That seems to me a most astounding mode of argumentation. Surely to be absolutely free for five years from these hospital diseases is a matter which would be of enormous moment in any hospital in this country. I do not hesitate to say that Professor Nussbaum has achieved splendid results, and all the more striking from the fact that his hospital was previously a pest-house.

Well, sir, I have been often reproached for not having published statistics, and it has been hinted, and the hint has been lately prominently repeated, that I have suppressed statistics because I had none which I should not be ashamed to produce.[1] Sir, the truth is that life is short, and that when every day begins one has to consider what is the occupation which is most likely to be valuable ; and feeling, as I do, very much as Mr. Macnamara does with regard to statistics, I have felt that there was every day something both more congenial and, I hoped, more profitable to do, than to compile statistics. But, recently, my friend Mr. Watson Cheyne, who has had my hospital books from 1871 open to him, has looked over my cases and made a statistical compilation, and I propose to-night to make some reference to what he has done.

I may say I do not think I am fairly open to the charge of having withheld statistics with regard to the question of hospital diseases. When I left Glasgow I published a statistical account with reference to those diseases, and showed that there had come over my wards, which previously had been amongst the most unhealthy in the kingdom, a wonderful change, a surprising alteration since the introduction of strict antiseptic treatment ; while in other wards of the same hospital the old conditions remained unchanged.[2] Then, at the time of the meeting of the British Medical Association in Edinburgh, in 1875, in the address I delivered as President of the Surgical Section, I gave a further account to the same effect, showing that in Edinburgh, although 1 had been working under most unfavourable hygienic conditions, nevertheless still further progress had been made in hospital salubrity.[3] Therefore I do not think it is quite fair to say I have never given statistics. But it seems that something more detailed is desired ; that it is not sufficient to say I have only had

---

[1] Mr. Savory, in his address at Cork last August, quoted with approbation the following passage from Mr. Bryant. 'The publication of isolated cases, however good, proves nothing, whereas the withholding of the whole suggests much.'

[2] See the *Lancet*, January 1 and 8, 1870.

[3] See *British Medical Journal*, December 25, 1875.

1 case of pyæmia in five years in a great hospital, with an average of 60 patients under my charge, but it is wished that I should say I have had so many amputations with no pyæmia, so many excisions of the mamma with no pyæmia, and so forth. I confess I do not see the reason for this desire; for, if I could be supposed guilty of making a false statement of having had only 1 case of pyæmia in five years with 60 beds under my charge, I should be thought equally capable of compiling a false table of statistics. We have lately had presented to us by Mr. Savory a sample of statistics, which I suppose may be regarded as a very good one, ranging over a large area, from the great experience of St. Bartholomew's Hospital, where there have been in three years— 1876-7-8—2,862 cases of injury, and 1,235 major operations—a large mass certainly to deal with; and he tells us that the deaths from injury have been 7·47 per cent.; the deaths after operation 5·82 per cent.; that the deaths from blood-poisoning, including pyæmia, septicæmia, and erysipelas, had been, after injury, ·42 per cent., and after operation 1·44 per cent. These statistics show a noble amount of surgical success, and, I do not hesitate to say, not many years ago would have been considered almost incredible. But I would remark, in the first place, that Mr. Savory gives the result *en masse* of the practice of four surgeons. Now Mr. Savory himself, as Mr. Smith has remarked, uses some antiseptic means. For instance, he employs the catgut ligature, which certainly renders death from hæmorrhage and also death from blood-poisoning much less probable than it was before. He also uses other antiseptic means. Further, Mr. Savory includes the practice of the late lamented Mr. Callender. I once went through Mr. Callender's wards, and I found that he was practising a very careful antiseptic treatment. He did not use the spray, it is true—and for many years I did not use the spray myself—but with that exception Mr. Callender was practising a thorough antiseptic treatment. His cases of compound fracture were treated with pieces of lint steeped in carbolic oil, and painted every day with it, reminding me exactly of my own early cases of compound fracture. I do not mean to say that Mr. Callender's good results were all due to this antiseptic element. I admired the pains and the ingenuity with which he avoided movement in his wounds and so forth, but I cannot doubt that his success was largely due to antiseptic management.

The practice of Mr. Smith is also included in Mr. Savory's

statistics, and Mr. Smith in many cases adopts the very practice which I have recommended; therefore, I say we cannot regard these statistics, excellent as they are, as statistics of a hospital where no antiseptic treatment is adopted. I would add that, even if Mr. Savory had used no antiseptic means whatever, he would nevertheless have derived advantage from the antiseptic practice of his colleagues. I was informed lately by a surgeon of the great Lyons Hospital, which used to be almost a pest-house, that one of the surgeons of that institution still refuses to use the antiseptic treatment, but, as the surgeon told me, he benefits by the practice of his colleagues. It is natural it should be so, because the effect of strict antiseptic treatment by three surgeons and non-antiseptic by the fourth is simply to convert a large hospital into a small one with reference to the question of hospital-disease.

I will now refer to my own statistics from November, 1871, to August, 1877, when I left Edinburgh. And first I will speak of injuries. Of these I have but a small number to refer to compared with the great array of Bartholomew's. The fact is, the great Infirmary in Edinburgh, while it is a metropolitan hospital for Scotland for surgical disease, is not one to which very many injuries come, and the great majority of these are treated as out-patients. And I may remark that with antiseptic management we are able to treat as out-patients many injuries which formerly we should have felt bound to admit into hospital. Thus I have only 72 cases of injury to speak of in these five years and three-quarters. Nevertheless, they were all somewhat severe injuries. 33 were compound fractures, and 7 were wounds of joints conservatively treated. In these 72 cases of injury I had 4 deaths, which gives us 5·7 per cent. The Bartholomew statistics are, for cases of injury, 7·47 per cent., so that mine are a little better. None of these 72 cases died from blood-poisoning. Then we come to operations. All my operations recorded in the case-books were 845; of these 37 died, or 4·4 per cent. Now Mr. Savory includes in his table only major operations. The term 'major operations' is very vague and arbitrary. I therefore have thought it better to put down all my cases of operation, without excluding—as Mr. Savory has done—any group for any reason whatever. But Mr. Cheyne, in going over the operations, has considered that there were 120 which might be fairly called minor operations. The great majority of minor operation cases, I may say, have been treated as out-patients. Subtracting, then, these 120 minor operations, I have 725 major operations. Of the 120 minor operations not one died, and therefore by subtracting them I increase my percentage of mortality. We still have 37 deaths, but these now give 5·1 per cent., while the St. Bartholomew's statistics give 5·82 per cent.—somewhat greater, but not very much so, I confess.

I cannot help remarking how easy it would have been for me to manipulate these statistics a little, and to make the result look better. For example, I had several operations which I have included among the major operations, which were very minor in point of operative procedure; for instance, there were three cases of spina bifida in which I tried antiseptic drainage. The operation consisted of introducing with a needle one or more threads of horsehair, and one case of chronic hydrocephalus was treated in the same manner—a very minor operation, certainly; yet each one of these was followed by death. I have considered that, in consequence of the greatness of the interest involved, it was only right to regard these as major operations, and every case in which I have had a death I have included in the major operations. Now, if I had chosen to manipulate, and say that these were minor operations, although they happened to be fatal, I might have greatly reduced the percentage of my mortality. But I have done what I considered more honest, and, as aforesaid, my mortality is 5·1 per cent. for major operations, as compared with the 5·8 of the St. Bartholomew's statistics.

Now, as to the causes of death, we come to the great question of blood-poisoning. I had 6 deaths from blood-poisoning in my 725 operations, or ·82 per cent. The Bartholomew's percentage was 1·44 per cent., or nearly double mine.

The cases of death from blood-poisoning were 2 of pyæmia, 2 of septicæmia, and 2 of erysipelas. But these cases require to be further considered. There is a very weighty statement attributed to Morgagni, to this effect, '*Perpendendæ non numerandæ observationes*,' which may be rendered in English ' cases should be pondered, not numbered;' and if we are to derive any benefit from these statistics, we must look into details, as has been remarked by Mr. Holmes. Now of the various operations I performed some were, from the conditions of the case, capable of being performed antiseptically; while in others this was impossible; as, for instance, where putrid sinuses existed in the vicinity of joints, or in lithotomy, removal of the tongue, and so forth, where we operate on parts where septic material must inevitably be present. There are therefore certain operations which we may call antiseptic, and others septic operations. Now, if I divide my operations into these 2 groups, I find that the antiseptic operations were 553, and that of these 553 only 2 died of blood-poisoning; and then, when we look into those 2, we find that one was a case of septicæmia occurring where the mamma had been removed, and the axilla cleared out to the collar-bone, and I was aware that on one occasion, ten days after the operation, all having till then gone on well, the spray was carelessly withdrawn from the wound at the critical time when the drainage-tube was being removed from the axilla. The other was a case of erysipelas, also after removal of the breast—but this was the only case of death from erysipelas after an antiseptic operation. The septic operations were 292, being little more than half the number of the antiseptic ones; yet the deaths from blood-poisoning were four,[1] or four times as numerous in proportion. That seems to me very instructive.

Then, further, if I divide the time into two periods—the time before the Edinburgh meeting of the British Medical Association in 1875, and the time after—I find, as might have been expected, that the latter shows more favourable results. 1871 was the date of the introduction of the spray, and at first we were working comparatively under difficulties. But since 1875 the antiseptic treatment has been carried out more effectually. Accordingly, we find that while between 1871 and 1875 the percentage of the deaths after all operations was 4·7, from 1875 to 1877 the percentage was only 3·7—that is to say, out of 295 operations we had only 11 deaths.

Then, if I look at the question of blood-poisoning in the last two years, I find that out of those 295 operations, to which should be added a certain number of accidental wounds, I had only one death from blood-poisoning, and that one death from blood-poisoning was a case of pyæmia, where I had performed a plastic operation to make a new nose. I endeavoured to turn aside part of the

---

[1] Of these four cases two died from pyæmia (one an amputation of the penis, the other a plastic operation on the nose), one from septicæmia (a case of excision of the tongue), and one from erysipelas (a case of small abscess of the neck opened by the house surgeon without antiseptic precautions).

ascending process of the superior maxillary bone to serve as a support. I split the bone with pliers; and I was quite conscious I had made a mistake. That was a case in which antiseptic treatment was impossible in consequence of communication with the nasal cavity. The patient died after rigors; but although I made careful search, and dissected the bones concerned and the veins, there was no pus found in the cancellated tissue or in the veins leading from it. Nevertheless, there were abscesses both in the lungs and liver, and I freely admit that the case was one of pyæmia resulting from the operation. That, however, was the only death from blood-poisoning in two years with 295 operations of all sorts, septic as well as antiseptic, besides various injuries.

In order to do justice to my subject, it is absolutely necessary to go a little further into detail. During the five years and three quarters I have referred to, there were 80 major amputations, and out of these I had 9 deaths, or 11·25 per cent. Now that, compared with what Mr. Erichsen says in his book on Hospitalism (published in 1847), that we must expect in major amputations from 35 to 50 per cent. of mortality, seems very good. But I confess I should not be at all satisfied with 9 deaths in 80 amputations without something more to explain them. Now, if we look into the details, first we had 3 amputations at the hip-joint. One of these was a primary amputation, and the patient was moribund on admission. I operated without practically any hope of saving the man, but I have seen such a thing as a patient coming round after being pulseless and in an apparently hopeless condition. Two other cases were not primary. One was the case of a tumour reaching considerably above Poupart's ligament. The operation was one of tremendous difficulty, and the patient sank as the immediate result of the operation. Such a case as that has no bearing whatever on the question of mortality in amputation as influenced by the after-treatment. The third case was one in which I operated for myeloid disease of the thigh-bone. At first I amputated through the trochanters, where the bone looked sound when divided; but afterwards, on making a careful section and a microscopical examination, the disease appeared to have extended higher. I therefore proceeded to amputate at the hip-joint the next day. The two operations, one following after the other, reduced that man so much that within twenty-four hours he was dead.

Now I say you may eliminate those three cases, for they do not bear upon the subject. The same is to be said of two deaths after primary amputation in the thigh. One was a double amputation. Both were in a state of collapse before the operation, and neither rallied. I had four primary amputations at the shoulder-joint, and one of the patients died. The case was one of railway injury, with the scapula as well as the arm extensively shattered, and the man was in a state of collapse when he was admitted, and never rallied. That case is in the same category—it has no reference to the question of mortality after amputation. We had one death from amputation at the shoulder for disease. It was a case of malignant tumour of the arm. The stump was doing perfectly well, but after some time the patient died of hæmorrhage from another tumour in the femur, of the existence of which we were not aware. This gave way, and bleeding took place into the substance of the thigh to such an extent that the patient died of the internal hæmorrhage. That case, therefore, had nothing to do with the question of amputation of the shoulder-joint. I had 25 amputations for disease in the thigh. One of these patients died, but it was nine weeks after the operation, when the stump was cicatrised, and he died from diphtheria. As far as the amputation was concerned, it was a case of recovery. I had 18 amputations of the ankle, and of these 1 patient died, but that was a boy, who three months after the operation, when the wound was almost absolutely healed, died of cerebral hæmorrhage. With regard to the amputation, this also was a case of recovery. Looking, then, at these cases of amputation, I may say that no patient died from a preventable cause; every patient recovered who had a chance of recovery.

Then there was one other class of cases which we used to consider exceedingly dangerous with regard to the risk of pyæmia— cases of ununited fracture treated by cutting down on the seat of fracture and removing the ends of the fragments. We used to operate thus upon some of these cases in the upper limb, but in the lower we considered that the risk of pyæmia was so great as to prohibit the procedure. In the five and three-quarter years I have referred to I have operated 8 times for ununited fracture of the thigh, 9 times for ununited fracture of the leg, once for ununited fracture of the clavicle, 4 times for ununited fracture of the humerus, and 5 times for ununited fracture of the forearm—

in all, 26 operations, and so far as I know all those patients are alive and well—not one of them died.

I would now refer for a moment to the statistics to which Mr. Bryant has so disparagingly alluded—the statistics of wounds of joints published last week by Mr. Cheyne. I will not dwell upon the cases of injury, because injuries are always uncertain subjects for antiseptic treatment; but I do say that a series of 20 cases of healthy joints opened, and kept open for days together, without a single failure as regards the septic element, is a fact of very great importance.

But here, Sir, we have an example of statistics of a totally different order from those which I have hitherto considered; statistics affording evidence of a new principle coming into play. I may be wrong, but it seems to me that, if you were to open a healthy joint, keep the wound open and put a drainage-tube into it, and take that drainage-tube out every day and wash it, and put it in again, not using antiseptic means of some sort or other, you would infallibly have more or less of inflammatory disturbance, and it would be impossible to have the condition of things which we now look upon as normal—namely, no tenderness, no blush of redness, no puffiness, no uneasiness, not to mention no suppuration. As far as I am able to judge, this is a kind of fact of a new order, showing that we have a new principle at work.

It has always seemed to me, therefore, of great importance to publish cases of this kind, even though they be only individual cases, which have been somewhat hardly reflected upon by some who have commented upon them. It has been said 'the publication of isolated cases, however good, proves nothing.' I say that one individual case, if it shows new pathological facts, is worth as much as a million. I have published, for instance, numerous cases to show that a great abscess connected with disease of the vertebræ may be opened by free incision, a drainage-tube introduced, strict antiseptic treatment being used, and that from that hour there may be not another drop of pus. I believe this is a fact as new to pathology as it is beautiful in practice.

Then I have pointed out again and again that with antiseptic treatment you may have a blood-clot exposed in an open wound under a moist dressing, this blood-clot remaining not only free from putrefaction, but indefinitely without suppuration, so that in the course of time, when you remove its upper surface, you find a scar beneath, without a drop of pus or a granulation having ever been formed. This, again, as far as I am able to judge, is new in the history of surgery, and indicates that we have a new principle at work.

Further, I have given evidence to show that dead tissue, if protected from putrefaction and also from the stimulation of antiseptic substances, may be absorbed in a way that I had never before supposed possible; that even a large piece of dead bone may have granulations grow over it, and these overlapping granulations may, as it were, eat it away till it is reduced to a very small size or disappears altogether. That, again, as far as I can judge, is an indication of a new principle.

Then, as regards the catgut ligature, that falls into the same category. Here we have a piece of dead tissue absorbed, and I have described how the process takes place—how there is an absorption of dead tissue, and how living tissue comes to take its place. That, I say, is something quite new. If these matters have not attracted attention, it cannot be because they are not worthy of it; I presume it is because I have not capacity to bring them before my professional brethren with sufficient force to impress them upon them. It is not, I say, that these things are unimportant; but that they are not believed.

With regard to the operation for transverse fracture of the patella, to which reference has been made by one of the speakers, instead of being regarded as a safe procedure, it was supposed to be a most dangerous operation, and hints were thrown out that if anything went wrong with the patient, I ought to be arrested on a charge of manslaughter. With regard to abscesses, one of our best provincial surgeons was recently at King's College, and saw some of the cases there. He said, 'Do you mean to tell me that in the case of a great psoas abscess connected with disease of the vertebræ no pus is formed from the time you squeeze out the original contents?' He spoke as if the thing had been only made out a day or two before, instead of being a matter of practice for years past. Then as regards the organisation of the blood-clot. I know that this has been simply discredited; but I will not occupy the time of this great meeting by going into evidence that this process really takes place.

Professor Maas, of Freiburg, lately sent me an introductory lecture which he had delivered to his students, in which he referred to the uselessness of statistics with regard to the antiseptic treatment. But he said he should like anybody who wants to be convinced of the value of the antiseptic treatment to try the following experiment, which has been often performed. In a certain number of rabbits tie the ureter with a silk thread without antiseptic precautions, and in a number of other rabbits tie the same organ antiseptically with catgut. Infallibly in those rabbits where the silk is used without antiseptic means you have suppurative pyelitis, while in the others you have simply hydronephrosis.

When at Amsterdam I had evidence of a similar character from two different and independent sources. One was the illustrious Virchow, who told me what antiseptic treatment had done for vivisection. 'Without the antiseptic treatment,' he said, 'if we took away an organ the inflammation and fever so complicated the matter that we could not judge of the results, but now with the antiseptic treatment we take away an organ and we can see uncomplicated the effects of its removal.' That seemed to me, stated in a very simple way, to be a beautiful and very important confirmation of antiseptic principles.

The other piece of confirmatory evidence which I obtained on that occasion was derived from two different ophthalmic surgeons present at the congress, who told me, that in using the antiseptic treatment in extraction of cataract, they found that instead of its being a question how much discharge there was after the operation, there was absolutely none ; and the sclerotic and the conjunctiva were as white as if the patient had not been touched. In other words, they had obtained results of an entirely novel character by the application of the antiseptic principle.

I feel I owe an apology to the meeting for having detained it so long, and I return you my sincere thanks for having listened to me so patiently. In such a gathering of medical men as I see before me I cannot avoid speaking warmly on a matter so near my heart. I have been charged with enthusiasm ; but I regard enthusiasm with reference to the avoidance of death, pain, and calamity to our fellow-creatures as a thing not at all to be ashamed of; for I feel this to be a matter of which I may say in the words of Horace :—

Æque pauperibus prodest, locupletibus æque,
Æque neglectum pueris senibusque nocebit.

The meeting, which lasted for upwards of two hours, was adjourned for a fortnight, on the motion of Mr. Jonathan Hutchinson, seconded by Mr. Hulke.

The adjourned discussion took place on Wednesday, December 17th, at St. Thomas's Hospital. There was a large attendance.

MR. JOHN WOOD, F.R.S., President of the Branch, occupied the chair, and, in opening the proceedings, said : Among the remarks made at our last debate, was a personal appeal to myself by Mr. Timothy Holmes. It was to the effect that he had, some time ago, suggested publicly that, now that Mr. Lister had come to King's College Hospital, there was offered a favourable opportunity of comparing the results of the antiseptic system of treating wounds with the more simple methods, in the same hospital and under the same hygienic conditions, and that he had been disappointed at my not having taken up the challenge and practised the simpler method exclusively in my own wards. I had practised antiseptic surgery according to my lights and opportunities ever since Lemaire, in 1860, established the value of carbolic acid in the treatment of open wounds. I saw at once a means of counteracting the contamination which even pure water, and clean lint and instruments acquired, when exposed in hospital wards. I had previously practised in all my cases Chassaignac's method of drainage by India-rubber tubes, and had arranged my incisions in all operations with a view to effectual drainage by gravitation.

When Professor Lister commenced the employment of carbolic acid, I looked with much interest for the results of his practice, which first appeared in a medical paper for March and July 1867. I had not a marked success beyond other plans with the carbolic putty treatment first employed by him ; but when he combined with the use of spray the antiseptic gauze dressing, with more satisfactory results, I instituted the plan in one of my wards, and started a Siegle's steam-spray and the gauze dressing for six months. My antiseptic bark (which was making very fair progress on the whole towards harbour) came suddenly to grief. It was among the shoals of expense that it suffered shipwreck. At that time, my wards were shared by the late Sir W. Fergusson;

and erysipelas having broken out among his patients, it spread to my side of the wards, the Listerian treatment notwithstanding.

The committee of management of King's College Hospital had spent much money in hospital arrangement and improved nursing, and an additional expense of 50*l.* in six months led to an intimation that, if I persisted in antiseptic experiments I must furnish the funds. This was not unnatural, for my surgical colleagues, and notably Sir W. Fergusson, were of opinion (still shared by many) that the pure waters of Damascus were as good or better than all the carbolised waters of Israel for purifying influence.

Before this time (following the note of preparation sounded by Mr. Erichsen), the full tide of hospital hygiene had reached and cleansed the London hospitals at a period long antecedent to its visit to the pest-house lazarettos of the Continent and elsewhere. This fact should not be forgotten in weighing the significance of the remarkable success of the simple treatment in St. Bartholomew's Hospital under Callender and Savory, whose statistics show only a fractional difference from those with which Mr. Lister has favoured us—a difference which may result from the fallacies lurking beneath all statistics.

Upon this tide came the wave of antiseptics, and topping the wave was the foamy crest of Listerism. Perhaps my metaphor will be more exact if I say the spray of Listerism. Of this wave, and of this spray, we took at King's College Hospital the fullest advantage.

Now it is rather too bad of Mr. Holmes to ask me to contravene all my cherished notions, to make all my patients the subjects of adverse experiment, and myself an awful example, for the purpose of furnishing forth a statistical table haply disastrous to some poor body, or even funereal. I have much reason to congratulate myself that I did not figure among the tables showing the very same thing given by Mr. Mac Cormac, and that I have thus escaped the oracular verdict of Mr. Holmes, that they were 'utterly and completely useless.' I agree with Mr. Holmes to a considerable extent as to the value of statistics compiled without regard to details of condition; and so far I find a further justification for declining to accept his challenge.

When Professor Lister joined the staff of King's College Hospital, he stipulated for and made arrangements which enabled him to obtain, complete seclusion of his wards, with a separate house-surgeon and nurses. One of my own wards freely communicates, and is nursed jointly, with that of another of my colleagues, and I have more than once had an invasion of erysipelas and other infections in that way. I think Mr. Holmes will admit that the conditions of the experiment could not, under these circumstances, be equal.

But I have tried to satisfy my own mind whether the system of Lister is so much superior to the more simple plan as to justify its increased cost. I have treated some of my cases by the simpler antiseptic method, and others by strict Listerism. I did so with the object of instructing my class in both, so that they might not be unprepared when opportunities came or failed.

Those who are destined for the army especially should know the simple method, for I am informed by army surgeons that it would be impossible to carry out the Listerian treatment in a campaign or after a battle. I have by the simpler method had some cases quite as good as any; tapping of knee-joints and injection of iodine for hydrops articuli, in a broken down constitution, without fever and with complete success; some cases of opened knee-joint, with drainage-tube through joint for three months, and recovery with a movable joint.

Some Listerian cases have shown signs of kidney-irritation, olive-coloured urine remaining undecomposed for a very long time; and other cases of skin-irritation (carbolic eczema). In one case, profuse salivation, lobular pneumonia, with olive-coloured urine, occurred after operation for removal of large myxoma from the face and head, with high temperature, getting better at once after removing the Listerian dressings. I have had one or two rare opportunities of comparing the two methods in the same patient, as to rapidity of healing, certainly on equal conditions. One was a case of removal of a benign tumour from each breast. The Listerian side was well ten days before the other. Whenever such opportunities have occurred, I have sedulously used them; and I am convinced of the exceeding great value of the Listerian dressings, and have employed the method in all serious operation cases, especially in abdominal and hernial cases, where the sac is removed, and in removal of varicose venous tumours.

In large cases of scrotal hernia in which trusses are of no avail and the sac is much thickened, of great size, and sometimes presenting constrictions in its substance which are a source of great danger from strangulation—the spray and gauze dressing, with the cleanliness and freedom from contamination, putrefaction, and suppuration which it affords, has enabled me to extend materially the scope of benefits to be derived from the operation for the radical cure of hernia. In such cases I have removed the whole sac and sometimes coherent omentum through an incision in the scrotum two and a half inches long, stitched up the peritoneal orifice with a continuous suture of strong carbolised catgut, and then I have drawn together the tendinous sides of the hernial opening, with thick silver wire, to resist the tendency of the intestine to protrude and force through the catgut suture (which is too weak to support the strain when unaided by such support, and becomes speedily absorbed). A drainage tube carried through the bottom of the wound and along the wire enabled me to keep the wound perfectly clear of accumulation and retention of discharges. The results I have obtained in the large number of cases I have done have been very satisfactory. In no instance have the symptoms been serious or even worthy of remark.

In several cases of large varicose tumours connected with the saphena veins, I have dissected out the whole mass under the spray, and tied the cut ends of the veins with carbolised catgut, with the result of perfect union without pain, suppuration, or any elevation of temperature whatever. In several cases I have removed two of such large clusters at the same time. One formidable operation in which the Listerian plan was carefully followed throughout is worthy of note. I removed the whole of a very large thyroid gland (which had undergone cystic degeneration), on November 10, 1877, from a woman aged twenty-eight. The case made an excellent recovery; the wound in this case was of great size and involved the ligature of many large veins and dealt with other important deep structures at the roots of the neck. After the operation there was one little suppuration, and the temperature although it rose once on the second day to 102°, remained at 100° for a few days longer and then gradually subsided to the normal standard. There is hardly any operation which could have tested better the value of antiseptic treatment than this.

I estimate highly the spray and gauze treatment in the opening of psoas, spinal, or other large abscesses for the first time. It enables us to keep the wound free from infective and putrefactive matters, and in the numerous class of cases in which the aspirator fails to empty the suppurating cavity of its inspissated contents it prevents the bad consequences of the access of contaminated air to the easily decomposed semi-solid contents.

In old cases presenting themselves as chronic sinuses leading deep into the tissues, the same advantages have not been evident. The spray and gauze treatment, however assiduously carried out, does not seem to cut short or to prevent the prolonged and lingering career of a disease so frequently hopeless. I have found as good results from the use of carbolic injections, drainage tubes, and the application of carbolised tow and cotton-wool dressings.

In many cases of chronic suppurating sinuses connected with diseased bone I have employed the internal administration of the sulpho-carbolates of iron and potass in doses of twenty to twenty-five grains three times a day with marked benefit. In these, as well as in cases where the spray and gauze dressing have been used for long periods I have found the urine, even though altogether free from carbolic discolouration, to be much less prone to decomposition when kept exposed to the air than it normally is. Now in diseases of a pyæmic type the blood and urine are prone to a more speedy decomposition than these fluids in a normal state; and the former is found also to be deficient in red corpuscular and in the inorganic salts, especially potass.

The plain inference is that a remedy for the disease must be looked for in this direction, if anywhere; and in the majority of the cases in which the olive-coloured change ever has occurred in the urine I have not found the patient's health to suffer from it, nor the appetite diminished. If pushed too far, however, the temperature rises and signs of carbolic poisoning appear.

Long-continuous observation of these cases has convinced me that we have the power to impress an antiseptic influence upon the secretions of the body generally, both by the external application and internal administration of carbolic acid, without endangering the health of the patient; and this I think shows that the exudations of a wound may be similarly and beneficially affected by such a course of treatment.

At a certain stage of healing, however, this effect becomes adverse instead of beneficial, and tends to prevent or prolong the process of cell-change in cicatrisation, if the treatment be persisted in. Hence the observations of practical surgeons employing the carbolic treatment have led a good many to recommend a change of dressing in the later stages of granulation and cicatrisation to one of a less stimulating character.

Little or nothing has been yet said in this debate about the bacterian theory of Pasteur as applied by Lister to the explanation of pyæmia, and the *modus operandi* of his method of dressing. It has been asserted that to practise his method rightly, you must be a disciple of the bacterian theory. As a prudential maxim, this is unexceptional, and may be useful in getting the work done properly. But, on looking over the past history of surgery, I find numerous instances where the practice was right, although the reasons for it were ridiculously and absurdly wrong. I will mention but one instance—the sympathetic powder of Sir Kenelm Digby. A simple dressing of clear cold water was applied to the wound, and the patient was all the better for it. The sympathetic powder was carefully applied to the weapon which inflicted the wound, and nobody was a penny the worse.

I have one or two difficulties to get over before I can adopt the bacterian theory. 1. How is it that, if bacteria have a power so terrible in their effects, in nine hundred and ninety-nine cases of wound out of a thousand they are unable to exercise it? 2. How is it that a patient may die of pyæmia or septicæmia, self-poisoned, without an external wound at all, the source of infection being a deep-seated abscess, far secluded from contact with the air? 3. How is it that bacteria may exist in abscesses originating inwardly, and yet no blood-poisoning ensue? 4. How is it that wounds of the face-cavities heal so quickly and so well, and yet bacteria are found in number in the fur of the tongue and the mucus on the surfaces of those cavities? It is, indeed, true that the mucus or saliva is antiseptic, but the bacteria are lively notwithstanding. It may be that the antiseptic properties of the saliva may prevent the bacteria from propagating, and prevent, in fact, the formation of the *zooglœa* which pathological experiments have proved to be so fatal when introduced into the system; and that the supply of bacteria is kept up by fresh relays from the atmosphere; or their liveliness may even be the expression of social discomfort and dissatisfaction, instead of that of joyous contentment.

But some admit to the full the bacterian theory, and yet deny the efficacy of the Listerian method. I cannot understand this. It seems to me that, if this bacterian pathology be true, Mr.

Lister's spray-treatment is its exact therapeutic equivalent. And I cannot admire the strategy which would assail Mr. Lister's position on this point, where I believe him to be impregnable. If clouds of bacteria hover in the air, seeking whom they may devour, we surely can best fight them in the air with clouds of antiseptic spray, and prevent them from coming to close quarters.

While thus defining the limits of my agreement with my esteemed colleague Professor Lister, I must take this opportunity of congratulating him sincerely upon the possession of those advantages which have made him so powerful an advocate of antiseptic surgery, and will give him so high a niche in the temple of scientific fame; upon the professional position, which has given him the authority; upon the gifts of fortune, which gave him the means; upon the gifts of nature, which gave him, in happy combination,

> The patient thought, the steadfast will,
> Resolve and foresight, strength and skill,

which he has laid upon the altar of suffering humanity.

Mr. JONATHAN HUTCHINSON: The admirable paper with which Mr. Mac Cormac opened the discussion, and Mr. Lister's speech, comprising his statement of statistical results, are documents of very great value, and, together with the contributions of facts and opinions from other distinguished surgeons who have spoken, must do much to help us to sound conclusions. Yet the subject still remains one of great intricacy and difficulty; and, although there are points respecting which I feel able to hold clear and strong opinions, on the general question I must speak with diffidence. On this matter we are in great want of detailed statistics, and have a surplusage of those which are vague and untrustworthy. Hence much bewilderment.

I am referring to the practical merits of the spray and gauze method of dressing wounds, which we now know as Listerism or 'the antiseptic method.' Respecting the antiseptic principle and the usefulness of antiseptics, we are, I think, agreed; it is as to the value, the comparative value, of that special method that some still doubt. As the time at my disposal is short, I may be permitted briefly to avow my own faith. I believe confidently that the spray and gauze plan of dressing is a very efficient plan of preventing the putrefaction of fluids in wounds; and that this prevention is of the utmost importance in avoiding wound-fever and blood-poisoning. I believe, further, that it will not only prevent the influence of common air as a cause of the putrefaction of dead material, but that it is also potent to kill the germs by which are induced certain more or less specific forms of inflammation, such as suppurative phlebitis, erysipelas, and allied conditions. I cannot doubt that these measures keep wounds sweet, and do this under conditions which are not favourable, and when, without them, these results would not have been attained. I have seen large abscesses opened under the spray and dressed with the gauze, which did not suppurate further, and which behaved in a way wholly unexampled under the older methods of dressing. I have also seen many large operation wounds so treated heal without suppuration, and without the slightest approach to a febrile condition. I have not yet ventured on any of the more daring procedures, such as opening comparatively healthy joints, or cutting into hernial sacs, with a view to radical cure; but I accept the evidence which has been recorded on these subjects. Let me be permitted to add, that I also quite believe that the details of Mr. Lister's plan are essential to its success.

There remain several explanations which I wish to make. I desire to make a few comments on the stages by which our modern opinions have been reached, and in explanation of the fact that those who do not employ the spray and gauze method are able to show that their results, in the gross, are nearly as good as if they did. My surgical memory easily goes back to the earliest periods of the movement which has resulted in a splendid reform of operative surgery, which has made the treatment of wounds comparatively without danger, and which has opened out to the surgeon many paths which were before impassable. It was not in one mind, but in many, that a zeal for the greater success of operations and a feeling of great dissatisfaction with their former results arose. Foremost, from the energy with which he devoted himself to the work, I must mention Sir James Simpson. He never tired of telling hospital surgeons that they were losing their patients on a scale which was far beyond what was inevitable, or of endeavouring to devise expedients which should help. In the belief that ligatures, and the bits of sloughing tissue which they were supposed to cause, were the means of poisoning wounds, an immense amount of ingenuity was devoted to perfecting the details of acupressure. The evils of hospitalism were loudly proclaimed, and the all-important duty of isolation of contagious cases was insisted upon. Next came the discovery of carbolic acid and its introduction into practice, and its use as a disinfectant of wounds. Earlier than this, however, Mr. Spencer Wells became editor of the 'Medical Times and Gazette,' and he at once devised a scheme for bringing to the light the real facts as to the mortality of operations. I was then his junior coadjutor, and the collecting of materials devolved upon me, but the plan was his. The design was by the publication of all operations in all the hospitals to which we could get access, to obtain data which should be trustworthy as to the rate of mortality of each, and the real causes of death.

Of these statistics, Sir James Simpson and others largely availed themselves, and I believe they were indirectly the cause of much of our subsequent gains. If my memory do not deceive me, the first nine ovariotomies which we thus dragged to light were all fatal. The ratio of improvement in the results of other operations has not been so great as in this one, but it has been very large. It was whilst our minds were much occupied with acupressure, torsion, &c., that we heard of disinfecting the air and the wound by carbolic acid. By degrees, Mr. Lister developed his practice in detail, and with steady patience kept close to his germ theory. The splendid results we partly know. But, during the development of Listerism, other and very important hospital reforms were in progress. Gradually every hospital was provided with an isolation-ward, and cases of erysipelas, gangrene, and the like were promptly removed from proximity with the healthy. Torsion of arteries and the catgut ligature superseded the old silk, and the proved value of carbolic lotions, &c., led to many other antiseptics —chloride of zinc, spirits of wine, iodine, terebene, &c.—being largely employed. My object in these statements has been to prepare the way for an answer to the question, How is it, if the spray and gauze plan of dressing be so valuable, that some of those who do not use it get such excellent results ?

The statistics of St. Bartholomew's have been brought forward, first by Mr. Callender and more recently by Mr. Savory, in proof that Listerism is, at any rate, not essential. Mr. Bryant has mentioned his own results in support of the same position, and I have often referred to my own. At the London Hospital, for many years Lister's precautions have been most ably and zealously used by one of my colleagues, yet his mortality has never been lower than my own. The explanation is, I think, to be found by reference to several facts. First, most of us (I speak now of those not using the spray and gauze) have been sedulous in the employment of some other antiseptic. Mr. Savory and Mr. Callender have used carbolic oil; Mr. Bryant, iodine and terebene; myself, spirits of wine and lead; and to these much of our results have probably been due. Secondly, I think an exaggerated impression as to the injurious effects of common air has got about. It was air laden with specific germs—those of erysipelas, hospital gangrene, and pyæmia—which in former times was so hurtful. From the risk of these, the isolation-ward and the frequent use of disinfectants for sponges, hands and dressings, have to a large extent freed us. Thirdly, it is quite possible that the spray and gauze plan, so useful in some cases, may in others be the cause of harm. Lastly, it ought probably to be admitted that those surgeons who show an unusually small percentage of mortality, whilst neglecting special antiseptics, are mostly of the cautious class, and abstain from operations involving peculiar risk.

One of the duties which must be undertaken in the future, will be to investigate the mortality under different plans of treatment of each several operation. Let us inquire as regards excisions of the breast, removals of loose cartilages from joints, amputations in each several part, excisions of each several joint, &c., and ascertain what the average mortality is under treatment by cleanliness without special disinfectants, and that under each kind of antiseptic application. The spray and gauze method does not appear to be well adapted for cases in which frequent movement of the part is likely to be hurtful—such, for instance, as compound fractures. Nor is it helpful where it is of great consequence to avoid a conspicuous scar. For the excision of tumours from exposed parts, I have had far better results in the way of immediate union and inconspicuous scar from the lead and spirit plan. For wounds which have already been exposed to risk of infection, I think that other means are usually preferable; and so also in all cases in which inflammatory conditions have already set in. It does not tend to repress inflammation, but only to prevent it; whereas the lead and spirit lotion does both. For most cases in which the surgeon makes his wound, the Listerian plan is admirable; and under this head we comprise all openings of joints and serous cavities, all osteotomies, most excisions of tumours, all openings of abscesses, and most amputations. So great are its advantages in this large department of surgery in diminishing suffering and preventing the risk of death, that there can, I think, be but one opinion as to the gratitude due from mankind to the genius, perseverance, and enthusiasm to which we owe it.

SIR JAMES PAGET:—I am not able to discuss the difference to which Mr. Hutchinson has referred between the complete antiseptic treatment with gauze and spray, and the less complete treatment which many have adopted. I must speak more broadly of the whole subject; and, for the sake of simplicity and brevity, I will speak of the treatments of only such wounds as are made in operations, omitting all mention of accidental wounds, compound fractures, and the like. Moreover, from among operations I shall exclude all such as tracheotomy and those for hernia, after which, if death follows, it is generally the consequence not of the wound, but of the continued disease; and from the first comparison of results I shall exclude ovariotomy, incisions of joints, and a few more, of which, however, I hope to speak before ending. And for nearly all cases I shall take my observations from my own practice and from what I have seen at St. Bartholomew's Hospital.

The reports which we have heard of the use of the complete antiseptic treatment in certain foreign hospitals make it impossible to doubt that it is absolutely potent for the repression of nearly all the fatal influences of foul air and the infective diseases of wounds. We have it stated on good authority that, speaking roughly, every wounded person brought into a certain hospital, or every person wounded there, became the subject of some infective disease, whether hospital-gangrene, or pyæmia, or some other; and that since the introduction of the complete antiseptic treatment that condition of things has ceased or has been reduced to a minimum. I do not doubt this, or try to depreciate so admirable a fact; but the question remains whether the same good results can be attained by other and, for the general conditions of both hospitals and private practice, better means. Now at St. Bartholomew's first Mr. Callender and then Mr. Savory have shown that the mortality of operations, with a partial application of antiseptic treatment, is as low as has been reached anywhere. This statement I can confirm; for the statistics first published

G

by Mr. Callender included the last three years of my practice in the same wards as he had when he succeeded me. And the hospital-statistics are confirmed by those of my private practice; for in respect of mortality the differences between hospital and private practice are very small. For many years I kept my papers separate, and then compared them; and the result showed that, with the exception of such differences as might fairly be assigned to the different characters and social conditions of the patients, there was no difference in the consequences of similar operations performed in a hospital or in private houses by the same person, at the same times, and with the same general principles of management.

In both hospital and private practice the mortality after operations has greatly changed during the time in which I have observed it; and it is useful to observe that the change has been, on the whole, a constant diminution. There has been no sudden change, as if by a suddenly introduced remedy, but, on the whole, a gradual diminution. I have been observing operations, and their consequences, for rather more than forty years; but I will take for comparison the period of thirty years, beginning at 1847, when I was appointed assistant-surgeon to St. Bartholomew's, and ending in 1877, when I left-off operating. During the first twenty-three of these years I performed and saw operations in both hospital and private practice, in the last seven in private practice alone.

In using numbers I must now speak roughly, and without pretence of having always had accurate statistics, such as, in this question, are certainly not yet attained or, I believe, possible. But, from some statistics and from strong general impression, I believe it may be stated that in the first part of the thirty years, say in the first ten years, the total mortality from all operations, capital and minor together, was not less than 15 per cent., that in the second period it fell to less than 10 per cent.; and in the third to less than 5. In the last seven years, when I had only operations in private practice, it was less than 2 per cent. from all sources. I think it certain that in thirty years the mortality, from all causes, after operations of all kinds, with the exceptions I have referred to, has fallen from about 15 to less than 5 per cent.

And if I take the larger operations, such as the larger amputations, excisions of the breast and the rest, I believe I may say that the total mortality has gradually fallen from more than 20 per cent. in the first period of the thirty years to less than 10 per cent. in the last. Let me repeat, I do not pretend to exactness in these numbers; there are good reasons why exactness cannot yet be attained; but they may serve to express with sufficient accuracy the diminution of mortality after operations at St. Bartholomew's and in my own practice, and, I believe, at many similar hospitals, and in the practice of other surgeons.

How has this result been obtained? The reports from some hospitals abroad show that a similar but even more striking result has quickly followed the introduction of antiseptic surgery, and that no other influence than this has been at work in them. I suspect that the belief in the absence of all other good influences is not quite just; but certainly in this country we have no facts that can be compared with theirs, and even those amongst us who are most devoted to antiseptic treatment can hardly doubt that the good results they have obtained may have been due, in some measure, to other less observed good influences.

Let me point out what have been the chief good influences brought to bear on the consequences of operations during the last thirty years, especially on those consequences against which the antiseptic treatment is directed. But, first, as I am referring at present to mortalities alone, it is to be observed that, in this country and in nearly all our hospitals, a large proportion of the patients are not susceptible of the fatal influences of the infective diseases. Whether this is to be ascribed to the power of resistance possessed by the patients or to a defective quantity or intensity of the infective material, or to both, I need not now consider. The fact is, that in the worst times for operative surgery that I have known a large percentage, 60, 70, or more even, of those who underwent capital operations, escaped death, and, I think I may say, escaped erysipelas, pyæmia, and the other perils against which antiseptics and our other methods of protection are directed. This large percentage, therefore,—this number of insusceptible or, as I may call them, not poisonable people, must be left out of account in our estimates of the comparative utilities of modern methods of protection. They escape now as they did then. When, with our present wise dread of putrid and infectious matter, we look back at the treatment some of them received, and the manner of dressing their wounds, it may seem strange that any should have escaped. Every stump, and every large wound at its first dressing diffused an abominable stench from the foul-smelling fluids shut up in it; yet the patients were not septicæmic. Patients with pyæmia, erysipelas, and putrid wounds were mingled in the same ward; yet they were not all infected; a large majority did well.

Happily a much larger majority do well now. Let me tell what seem to me the reasons for this improvement even where the antiseptic treatment is not employed, or is employed in a very limited manner.

First, during the last thirty years surgeons have left off some bad practices. At the beginning of this period the practice of bleeding before or after operations had not ceased. Some still thought it necessary to prepare certain patients for operation by taking blood from them; and more were in the habit of applying even a large number of leeches over the parts near wounds when any flush of erysipelas or of deeper cellular inflammation appeared. Still more general was the custom of giving active aperients, salines, antimony, and other medicines which may, at least, have done harm by interference with useful quietude. Some surgeons preferred an extremely low diet for even many days after an operation, a few thought it useful to give, even from the first, large quantities of stimulants. All this, I suppose, has passed away; for ten or more years there has been, in the place of these things, a wise simplicity of after-treatment; and I cannot doubt that it may be credited with a share in the diminution of mortality.

Again, in the last twenty years we have had vast changes made for sanitary purposes in all our hospitals, and a quantity of sanitary work past counting done in our large towns and houses. The tens of thousands of pounds which I have seen spent for this purpose at St. Bartholomew's have told of only a large instance of the care taken everywhere to exclude or weaken the sources of hospital-disease; and if we are right in believing that sanitary science has in any way diminished mortalities, we cannot doubt its having had some influence on the mortalities of patients after operations. They are the very class on whom pure air and water, drainage and ventilation, isolation, and every form of cleanliness, would have most influence.

During this same period of about twenty years the system of careful education of nurses has been introduced. There were always, among the sisters of our hospitals, some nurses excellent in every way; but some of even this class were ignorant and careless; and among the inferior nurses skill was not looked for and care was not usual. It would be difficult to exaggerate the contrast between the nurses of the first five and of the last five years of the period I am speaking of, contrasts equally great in private and in hospital nurses. It seems to me impossible to doubt that the contrast is shown in some measure of the diminished mortality of the operations.

Again,—medical education has in thirty years improved; and, among its results I think that all operating surgeons must have felt that they have had better help in the care of their patients, not only from their house surgeons, but from their personal assistants and from the general practitioners who were in more constant attendance than themselves. This cannot have been without influence. I believe it has been at least as useful as the improvements in method and in skill which time has brought in nearly all operations. Counting both together—the better operations and the better assistance after them—we surely may believe that in thirty years' progress they must have diminished our mortalities.

Lastly, during the last ten or fifteen years, there has been among surgeons a constantly increasing rivalry for the attainment of the greatest possible success after operations. All, I think, have felt that, if any cases should do well under them, their operation cases should. There were always a few who by constant personal care were distinguished for success; but the care which was singular has become general: and when I think of the amount of thought and attention which most surgeons gave to patients after operations thirty years ago, and the amount given by the same surgeons or by others in the same positions now, it seems as if here were cause enough for nearly all the diminution of mortality which I have observed.

Here, then, are five, and there may be more, influences for good which, in the last thirty years, have been brought to bear on our patients after operations. Of course, no one can tell how much each of them has contributed to the diminution of mortality which has been in the same time effected, but it would be unreasonable to deny to them a large share in it. I believe that to them may be assigned by far the largest share of the diminished mortality at St. Bartholomew's and in my own practice. Say that after operations, in conditions of which we should now be ashamed,

85 per cent. of the patients escaped and that, in present conditions, more than 75 per cent., escape, or that, after capital operations the mortality has been diminished by 20 per cent: surely the changes I have indicated may explain nearly all the difference. I do not in the least depreciate the value of the antiseptic treatment; I do not doubt that it has achieved the success assigned to it in hospitals less well managed than our own; but, so far as the mortality after most operations is concerned, I believe that equally good results have been and still may be obtained without it, or with a very partial use of it. At least, it seems clear that with care and watchfulness and scrupulous cleanliness in well-managed hospitals and private houses, there is little left for the complete antiseptic treatment to do. It may be that some day results yet nearer to perfection will be attained by the union of complete antiseptic surgery with complete sanitary management. But I hope there will be no attempt to prove in this country that antiseptics are self-sufficient even when there are neither good sanitary arrangements nor skilled nurses nor very watchful surgeons.

I have spoken only of the mortalities after operations; but a mere escape from death is not all we have to wish for. It is not enough that a patient should just escape with his life; it is desirable that he should not be in danger from fever or from erysipelas, or any other acute disease, nor yet from chronic pyæmia or long-continued suppuration, or any other malady introduced or aggravated by the operation. He should be cured with as little illness as possible, and with as little increase as possible of the tendencies to tuberculosis or any of the degeneration of internal organs, which may be associated with the disease for which the operation is done.

Our statistics tell only that certain persons either did or did not die; they ought to tell much more than this. I kept for many years a set of notes which indicated not merely the mortalities after my operations, but the well-doing or the ill-doing of each patient; and the observations which I thus made leave me in no doubt that with even a partial antiseptic treatment the recovery of patients after operations was quicker and more free from constitutional disturbance than when I did not employ it. During the last twelve or more years I always washed wounds with the 40-grain solution of chloride of zinc. I rarely used the carbolic spray, much more rarely any drainage tubes. I always used either torsion or carbolised catgut ligatures, closed wounds as exactly as was possible with silver sutures and plaster, deferred as long as possible the first dressings, gave moderate quantities of food, and very rarely any kind of medicine. I believe that by even this measure of antiseptic treatment my mortalities were diminished; but the influence of antiseptics could not be separated from those of the other improvements which I have mentioned. I more than believe—I am sure, that the recoveries were quicker, more direct, more free from risks of septicæmia and of the aggravation of chronic diseases than in the earlier periods of my practice, in which I used no antiseptic means, and less simplicity of dressing.

In all that I have said I have had in mind only the cases which occur in what may be called the general run of operative surgery. But there are some groups of cases in which I believe it would be absolutely wrong to dispense with any of the precautions of the complete antiseptic surgery. Among these are the cases of ovariotomy. I look back with remorse to my experience in them at St. Bartholomew's, when I compare it with that of Mr. Smith. All my cases were in the first half of my thirty years' practice, and no doubt part of his greater success may be ascribed to the better nursing, and better hospital arrangements, and better plans of operating which the last fifteen years have supplied. But whatever may be ascribed to all these, I am as sure as he is that the thorough antiseptic treatment has largely contributed to the success which he has attained. His success contrasted with my failures at St. Bartholomew's strongly confirms the experience of Mr. Spencer Wells, Dr. Keith, and Mr. Thornton. Their success without antiseptics had, indeed, been so great that it is hard to estimate the increase of their successes with them; but their general impressions on the question are as decisive as any statistical facts relating to it that can be attained.

As with ovariotomy and (I may add) with all abdominal sections, so with osteotomy and the cutting into healthy joints. I cannot doubt that operations of this kind, which, in the earlier years of my work were done with great risk or, with a wise fear of the risk, were left undone, may now, with antiseptic help, be done with an almost complete safety. In this direction antiseptic treatment has certainly enlarged the range of useful and safe surgery.

And another group of cases in which it seems to be of the highest value is that of large abscesses. A few years ago I believed that I had never seen a patient recover after the opening of a lumbar or a psoas abscess with a free incision; I could not remember one who had not died before the opened abscess healed. Of late years I have known such abscesses opened with complete impunity under antiseptic treatment, and there has seemed nothing but this treatment to account for the difference of the results.

Let me briefly sum up the opinions I have formed on antiseptic surgery.

I believe that, in its complete form, we can nearly neutralise the evil influences of unhealthy hospitals and other like sources of those infective diseases from which arise the largest portions of mortalities after operations.

That it has not yet reduced the death-rate to a lower level than can be attained by good sanitary arrangements, good nursing, strict care and cleanliness, quietude, and simple dressing.

That recoveries after operations are quicker and more free from fever and other constitutional disturbances when antiseptics are used than when they are not used.

That in certain groups of cases, such as I have enumerated, operations may be safely done with antiseptics which without them would be very hazardous.

And now let me end by saying that of all the achievements of surgery during the last thirty years, I regard the diminution of the risks of operations as by far the most important; and that, beyond comparison, he who has contributed most to it is Mr. Lister. More than any he has done good both by his own work, and by provoking others to do their best in their own ways.

MR. LUND (Manchester) said :—I shall speak a little as a partisan, because I have had, from the very first, a great fondness for the antiseptic system. From the first, I saw that there was a principle underlying it. I had, at a very early stage of Professor Lister's operations, gone to Glasgow for the purpose of observation, and I have tried to adopt the most recent alterations : and the general conclusion to which I have come is, that there is in the principle of antiseptic dressings an immense aid in the management of surgical cases. But the great question is, How far is it essential to carry out the principle of antisepticism in every case? What I see in the daily practice of that method is that it requires personal supervision; it can be entrusted to so few, it requires such constant attention, that cases are likely to go wrong. It might often be said that cases are treated antiseptically which are not so treated; but, where it can be carried out completely, the effects are most astonishing. As to trying to trace out how in some cases it succeeds, and in others it does not, I think an observation that fell from Mr. Hutchinson opens out the subject. In wounds made by surgeons—perhaps not amputation-wounds— wounds in small tumours or the opening of abscesses, how is it that it answers well in some cases and not in others? I think the answer is, that the antiseptic treatment is hopeless, unless it is begun early. We do not succeed with our larger operations, such as amputation of the thigh, because it is very difficult absolutely to shut out all sources of putrefaction. I think the great benefit of the more careful application of general antisepticism is the avoidance of early putrefactive change. I know the difficulties of carrying out the method; but I still think it is a most valuable addition to the practice of surgery; and could it be carried out in a way more generally applicable to everyday practice, and to the varying conditions under which patients are placed, the universal adoption of the system must inevitably result.

Dr. NEWMAN (Stamford) :—In 1871 I spent my summer holiday in Edinburgh, and gave much of each day to watching the details Mr. Lister's work in his wards. I came back from Edinburgh fully convinced of the value of the antiseptic method, and I have since carried it out very thoroughly in almost every instance in which it has been possible. Both in hospital and in private practice I am by conviction a thorough believer in Mr. Lister's plan.

The infirmary with which I am connected is small, and was built fifty years ago; the local conditions of the building are, even now, not of the best; the interior air space is deficient, and the sanitary arrangements are not by any means perfect. In the fifteen years of my practice there I have known, in proportion to the number of patients, a very serious prevalence of septic trouble more than once; much has been done to remedy the defects of the building, and these outbreaks have become less marked of late years. The more severe instances of blood-poisoning occurred, for the most part, prior to the time when I became familiar with the antiseptic treatment. But I lay stress on these unsatisfactory local conditions because I am very clear that these existed when I was obtaining a considerable measure of success in surgical cases treated on Mr. Lister's method.

For a time, at least, I stood alone in my belief in the value of this mode of dressing; and therefore the opportunity was offered every now and then, of having parallel cases in beds lying side by side, when not only I, but those also who were working with me, and seeing what was going on, could for themselves observe the influence of this special method as compared with that of the older plan.

I well remember that two compound fractures of the leg were admitted within a week of each other; one was under the care of a colleague, the other case under my own care. The one had not left the hospital at the end of six months; whilst the other, under antiseptic treatment, was discharged well at the end of two months. The fractures were almost precisely similar; the ages of the patients almost identical; both patients were agricultural labourers, and, so far as could be observed, they were both free from constitutional taint. The one left at the end of two months, with his fracture sound, and in all conditions as if he had met with a simple injury; the other, at the end of six or seven months, was still in the infirmary, recovering from deep-seated and huge abscesses.

I believe that I may claim to be the first surgeon in England to practise ovariotomy antiseptically. I did the operation in September 1871, and tied the pedicle with catgut ligature.

That ligature, I have no doubt, gave way at the end of four or five days, and some considerable hæmorrhage followed into the cavity of the abdomen. But the antiseptic treatment had been carried out thoroughly, and but the slightest constitutional disturbance happened. Yet, warned of the accident by the existence of local fulness and pain in the abdomen, that treatment allowed me in full confidence deliberately to break open the lower angle of the wound, and give exit to a quantity of effused blood. No kind of harm followed, and the patient recovered fast. The details of the case were published in the 'Medical Times and Gazette' of February 17, 1872.

It seems to me, however, that few men, comparatively speaking, realize how thoroughly little matters of detail are involved. As an instance in point, I would quote the following case.

A lady met with a very serious compound dislocation of the ankle, and I saw her, in consultation with a friend, who had read, but had not seen the details of Mr. Lister's method. We saw the patient a short time after the injury, carefully washed out the wound with a strong 1·20-solution of carbolic acid, and dressed it under the spray with the antiseptic gauze in the usual mode. To my horror, when I saw the patient four days afterwards, I found the wound suppurating freely beneath the protective and gauze: and no wonder, for part of the so-called antiseptic treatment had consisted in the deliberately putting upon the raw surface of the wound, beneath the protective, a rag dipped in a five-per-cent. solution of carbolic acid. That alone was sufficient to provoke abundant suppuration. It was an antiseptic treatment as regards the use of carbolic acid in gauze and spray ; but it was clear that the practitioner had not taken in how necessary it was to observe certain conditions.

Much has been said and written about the comparative mortality under this and differing plans of treatment : but I am very far from thinking that a question of this kind can be settled merely by a reference to the mortality. The entire absence of elevation of temperature, and the entire absence of constitutional disturbance, are great matters ; they add materially to the comfort of the patient, they diminish the time during which the process of repair must be continued ; and the patient, from the very first moment that the operation-wound is made, is put into a condition of comfort and safety that no other treatment—no other half-antiseptic treatment—can secure for him. I am glad, therefore, to declare myself, by conviction and practice, a thoroughly warm supporter of this special plan of antiseptic treatment.

MR. KNOWSLEY THORNTON said : When I first heard of the debate on Antiseptic Surgery, I at once carefully collected the published statistics of ovariotomy, because I felt that a comparison of the results obtained in such an operation, before the introduction of Lister's method and after, must have great weight in deciding for or against the method. The statistics brought forward in Mr. Mac Cormac's admirable opening address, and the remarks of Mr. Spencer Wells, must be so convincing to those open to conviction, that I do not intend to dwell on the figures I have collected ; especially since I agree to a great extent with what Mr.

Macnamara has said at the previous meeting as to statistics in general, and with what has just fallen from Sir James Paget on the necessity that details of the cases should accompany the statistical tables. A good example of this necessity is seen in the statistics of antiseptic ovariotomy, for cases of death from septicæmia appear and may be taken as evidence of the failure of the method, whereas they arise generally from the cyst contents being putrid before the performance of ovariotomy, the putridity being from previous tapping without antiseptic precautions.

There is one point in the statistics I have collected, to which I will refer. In more than 300 recorded deaths from ovariotomy, over one-third was put down as distinctly due to septicæmia, pyæmia, or septic peritonitis; and nearly another third to peritonitis. 'Peritonitis' as a cause of death, while frequent in the earlier tables, almost disappears in the later, and is replaced by 'Septic Peritonitis,' this fact being probably due to greater accuracy in recording cases and to a better knowledge of the septic diseases. Peritonitis is almost unknown in antiseptic ovariotomy, and putting these two facts together, it seems fair to claim the deaths recorded as due to peritonitis as of septic origin. Thus, two-thirds of the mortality is due to septic diseases, and these being eliminated by Listerism, the mortality of the operation in the hands of those who faithfully follow Lister's method has been remarkably reduced. Other causes of death frequently recorded in the tables, were exhaustion and obstructed intestine; and the great majority of these are also, in my experience, distinctly to be traced to septic mischief. The diminished mortality following the introduction of antiseptics in the practice of experienced ovariotomists, who had apparently reached their minimum before using Lister's method, is hardly in accord with the views of Sir James Paget; for in no other surgical procedure have such perfect hygienic measures been applied as in ovariotomy; and yet the addition of Lister's method has marvellously lowered the mortality in the hands of specialists, and has made fair success possible in general hospitals. When some years ago I came to London, fresh from the teaching and practice of Mr. Lister, I was an enthusiast in antiseptics; and what I have seen since has not abated my enthusiasm. I had the good fortune to come at once to the practice of Mr. Spencer Wells. I think that Mr. Wells, by showing what could be done with attention to detail to improve one great surgical operation, has done more than any living surgeon, short of perfect antiseptics, to improve general surgery ; and if his example had been more followed by surgeons generally even better results might have been obtained than those dwelt upon by Sir James Paget.

When I first saw the careful preparation of instruments and sponges, the brilliant and rapid operations and the careful nursing and personal care afterwards, I was much struck and almost doubted whether Listerism could improve matters. But as time went on, I saw that the majority of deaths after ovariotomy was due to preventable causes, the chief among these being septicæmia. My faith in antiseptics remained firm, and now it is proved to be a sound one.

There is, however, something more for the surgeon to aim at besides the recovery of the patient. I was much struck by remarks made to me some time back by Sir James Paget, as to the 'well-doing and ill-doing of cases.' I would claim then for perfect antiseptic or rather aseptic ovariotomy, 'well-doing,' beyond what was before attainable. The healing of the wound made in abdominal section should be a very simple affair. The tissues are lax and one can give the wound efficient support, and keep it at rest. But compare the process of healing under the old system with dry dressing, and under the gauze dressing. In the first case, it is true, the majority of the wounds healed by first intention, but there was a necessity for frequent change of dressing, a very serious matter as regarded the comfort and quiet of a nervous woman ; then, if the silk sutures were left more than four or five days, there were redness and tenderness round the points of insertion, and if, from abdominal distension, it was unsafe to remove them, this redness spread and pain, fever, and troublesome suppuration frequently resulted, and I have even seen abscess in the abdominal wall and death from pyæmia as a result ; now the same silk is used, but it is soaked in carbolic lotion, and all the other parts of the operation being strictly carried out on Lister's method, the gauze dressing may be left for a week unchanged, or even for longer ; and at the end of that time, when it is removed, the wound is found soundly healed without a particle of redness or irritation round either its edges or the sutures. Here, then, is perfect 'well-doing,' both for patient and surgeon. The one is saved discomfort, pain, and the fear of frequent dressing, with possible danger to life, and the other, both anxiety and trouble. I maintain that such results cannot be obtained by any other method. And if this is the result in the external wound, which we can see, we know that the same result is surely taking place in the internal wounds hidden from our sight, but whose progress is indicated to us by the pulse and temperature.

If, after antiseptic ovariotomy, all goes well for the two or three first days, the result is pretty certain, and one visit a day will be sufficient where three or four were formerly necessary; because, under the old system, a patient apparently doing well would suddenly get a rise of temperature, and show septic symptoms, and the period of their onset was uncertain even for a week or more after the operation.

I think, when the future history of surgery comes to be written, the saving of anxiety in great operation cases will be by no means the least part of the credit which will be given to Mr. Lister. Surgeons will be ever grateful to him for freedom from the great anxiety and uncertainty as to when an operation case is safe from septic mischief; and patients, besides the priceless boon of immunity from those terrible and uncertain foes, and the smaller comforts already described, will have to thank him for enabling their surgeons to give more perfect consideration and study to each new case as it presents itself, because their attention will not be distracted by the 'ill-doing' of other cases.

Mr. MORRANT BAKER: If using carbolic acid and other antiseptics freely be sufficient to make one an antiseptic surgeon, I would claim to be one in the fullest sense of the term.

But it is not with a view to enter into the general question of antiseptics that I venture to take part in the present discussion, but merely with reference to the comparatively narrow question, Is it necessary in all cases of operation (when the method is applicable) to employ that mode of antiseptic treatment which has been introduced by Mr. Lister, and which Mr. Hutchinson calls the 'spray and gauze' system? To this question, I would answer 'No'; at the same time agreeing entirely with those who have stated in this debate that it would be, with our present knowledge, almost criminal not to employ Prof. Lister's system in certain cases. With reference to such cases, it seems to me that each surgeon may lay down a very simple rule for his guidance, by asking himself the question,—would he desire the method to be employed, if he were so unfortunate as to require the given operation to be performed on himself? If he would, he surely ought to employ it for others. But in many cases, the method appears to me unnecessary, or even harmful; and as a justification for stating this belief, I will quote a few statistics; remembering what has been said by a former speaker, that we can arrive at truth in this matter only by taking groups of cases, and comparing one group with another. The special group to which I shall now confine myself is that of capital amputations of the limbs.

The total number of capital amputations performed by myself is 52: of which 3 were amputations at the hip-joint; 15 of the thigh; 1 at the knee-joint; 12 of the leg, almost all in the upper third; 6 at the ankle-joint; 3 of the upper arm; and 12 of the forearm. Of these cases, only *one* died from a cause directly attributable to the operation, and *not one* from septicæmia or any allied disease. The patient who died was a man aged 47, whose leg was amputated in the upper third on account of an intractable ulcer. He was a degenerate subject altogether, and we could never get his stump to heal. Nearly 5 weeks after the operation, when he died, the flaps looked very much as they might be expected to look a few days after the amputation; and at the post-mortem examination, the only abnormal conditions found were slight interstitial nephritis and a large and very fatty liver. He had not at any time symptoms of septicæmia.

Of the remaining 51 cases, 5 died during residence in the Hospital; 2 of them within an hour or two of the operation, from shock produced by other injuries than those for which amputation was performed; (in both of these cases the amputation was deferred for a while, lest the patient should die on the operating-table). The remaining 3 cases died at periods respectively of 3 weeks, 5-6 weeks, and 6-7 weeks after the amputation, from advanced phthisis or lardaceous disease. In all these 3 cases the wound was practically healed; in one of them so completely that not even a sinus was left. In none of the 52 cases was the 'spray and gauze' system employed; and the question which seems to me to fairly arise from their consideration, is this,—would an equally good result (viz. *one* death, and that not from septicæmia, in 52 cases of capital amputation) have been obtained if this system had been adopted? And on the evidence now obtainable, I doubt if it would. Professor Lister's system has evils of its own; and, therefore, valuable as it is for such cases as ovariotomy or operations, say, for removal of a loose cartilage from a healthy joint, it may be harmful, if unnecessary, in others. I do not refer to the discomforts attendant on its employment. These must be reckoned as the merest trifles, if the method be good for the patient. But there are real evils unavoidably attendant on it. Carbolic acid poisoning, though usually harmless in the end, is not by any means always so. I have seen cases in which the patient has seemed in great peril from it, and one case at least in which I could not but suspect it as a cause of death. I have seen a case, (under the care of a surgeon thoroughly familiar with all the details of the system, and who had employed it for many years) in which the infiltration of a limb with pus, necessitating amputation, has appeared to be the result of the bandaging and confinement which the system entailed. There are some present, I believe, who could tell of serious or even fatal recurrent hæmorrhage which has been concealed under the gauze and bandages. And there *must* be a certain number of cases in which the system will fail; and if it fail, all would agree that it may be a source of great danger. It is no answer to the last-named objection, that the system ought not to fail. Fail it will, in a certain, though it may be a small number of cases. And if it be true that there are cases in which the system is *unnecessary*, then these cases of failure must be reckoned as so many objections to its employment. I would repeat then that while fully recognising the value of Professor Lister's method of treating wounds in many cases, and heartily joining in the expressions of gratitude which have fallen from so many, for the brilliant services he has rendered to 'antiseptic surgery,' I think the results I have mentioned in a certain group of cases justify me in the opinion which I have expressed. Time only will tell us in what cases the spray system should, and in what cases it should not be adopted; and we shall never arrive at the truth in this matter until we are prepared to discuss candidly what there is to be said on both sides of the question.

MR. MAC CORMAC, in reply, said: I venture to think that

this discussion is one of the most valuable contributions to surgery which has been made for some time. The conclusion to be drawn from the discussion, it strikes me, is, that every one of those surgeons who use antiseptic means desire to know what is the best method of using antiseptics, and what is the most easy and efficient method by which they may contribute to save human life and prevent suffering. That Mr. Lister's method does immensely contribute to these great ends, I think this discussion has fully and certainly established. Since Sir James Paget has given the challenge, I would mention one case which has recently come under my own observation.

On the 1st of this month, there came to this hospital a man of sixty-two years of age, an habitual drunkard, with his fatty tissues. He had been run over by a heavily laden van, two wheels of which passed over his leg, fracturing the right tibia and fibula, and extensively separating the skin from the subjacent fascia, and also the muscles from one another. This injury was treated by the resident assistant surgeon, who, for the space of an hour, took every pains to render the large wound aseptic, and employed all the means at his disposal, required for the completion of the first dressing in the most perfect manner. He made three large openings into the tissues of this limb, and removed two large portions of the tibia. The man had also a wound on the head. He had during the treatment a slight attack of delirium tremens, but beyond this he has gone on well; his limb has not shown any of the symptoms you usually expect, and the only question that presented itself to the mind of our resident assistant surgeon, Mr. Pitts, was whether he was justified in attempting to save the limb. At the end of sixteen days, the patient is doing well; he has no local pain, his wounds are not inflamed; there is no discharge of pus, no abscess. He has had, however, a slight attack of erysipelas in the head, connected with the injury there, but no erysipelas on his leg. He asks repeatedly for food; and has been doing in every way very well.[1]

I happened to read to-night a pamphlet by a surgeon connected with a hospital at Cologne, who mentions incidentally, in a paper discussing the question of removing malignant tumours from the rectum, what the results of his various operations were in 1878. He had performed 44 amputations, amongst them being 17 amputations of the thigh; 53 resections, 15 of the hip and 12 of the knee; 13 wedge-shaped removals of the bone, 10 amputations through the joints, one being a disarticulation of the hip-joint, 15 operations for false joints with 'refreshing,' as it is called, of the bone-ends: and not one single case out of this large number of important operations had any fatal or unfavourable result.

I am, of course, with the limited time at my disposal, unable to review the numerous facts and opinions which have been brought before us. I ought to state that my attempt to bring forward some statistics, which I did simply as a contribution for what they were worth, was treated with some contempt by Mr. Holmes, although I have brought forward amongst those tables some which I thought would have been sufficiently satisfying. He had asked for a comparative series of statistics showing the results that might be obtained in the same hospital by two surgeons—one practising antiseptic surgery, and the other not. I found that such an experiment had been made in the Glasgow Royal Infirmary. The results of the comparison showed that, whereas one surgeon during three years treated 1,884 cases non-antiseptically and had a mortality of 5·84 per cent., another surgeon treated 1,706 cases antiseptically and had a mortality of 2·93 per cent., or just half the other.

[1] The man some time after died from abscess in the brain. The compound fracture was aseptic throughout.

Other statistics were given in which I thought the special merits of the particular kind of treatment were well exemplified. Amongst these there were examples of success obtained in spite of the great and serious difficulties attending the treatment of wounded men belonging to an army actively engaged in war, and I claim to know something of what those difficulties are. I told of one surgeon who had been able to save many such serious and fatal injuries as gunshot wounds penetrating the knee-joint. And, although your Chairman has said he has been informed that it is impossible that antiseptic surgery should be applied in military operations, I can assure you that it will be done, that indeed it has been done; and that in Germany, where the surgeons have had a larger experience, perhaps, of late times than in other countries of injuries produced in war, the surgeons one and all of that country are striving, and already with some success, to establish the best means whereby antiseptic surgery shall be applied in time of war.

I have to thank you for the kind manner in which you received my Address, and also for the patience with which you have listened to me to-night.

ON THE ANTISEPTIC THEORY.

In a sense antiseptic surgery is as old as surgery itself. The balsams, the boiling oil, which Paré set aside, the vegetable and mineral acids, alcohol and many other remedies, recommended themselves empirically, to neutralise the ill consequences so commonly accruing after wounds, while the external temperature, the season of the year, a stagnant or impure atmosphere and other such things were observed to influence their progress, and more or less effort was being constantly made to guard against these influences.

Modern antiseptic treatment, however, differs from all that preceded it in this essential respect, that it recognises a definite specific cause for wound-inflammation, and attempts to keep away, or destroy, the elements in the air which it affirms produce decomposition in wounds, and certain diseases depending thereon. But besides this, much is needed to procure the satisfactory healing of a wound. To the method employed to prevent putrefaction must be superadded many auxiliary means which are the common property of all surgeons.

It would be clearly wrong to attribute all the improvement gained by antiseptic surgery to the use of antiseptics alone.

Greater pains are taken to still blood. Greater exactitude is likewise taken in bringing together the edges and surfaces of the wound, and a more careful provision made for the outflow of secretions. The studied avoidance of mechanical and chemical irritation, or a physiological rest, and the infrequent change of dressings also contribute their share. The importance of these matters led one by one to those improvements which have been introduced by Lister into his original method, as, for instance, the discovery of catgut for ligatures and sutures, the application of a protective material, the use of drainage, and the quite recently perfected form of drainage by means of Neuber's absorbable bone tubes. The absorbent character of the dressing is a further and great security, since the secretions at once become disinfected so soon as they escape into it.

The theoretical basis on which the antiseptic method rests remains unchanged through all these manifold changes in detail.

The value of Lister's, or of any other plan of treatment may be chiefly judged by two things—the mode in which wounds heal under it, and the influence it possesses to ward off accidental wound diseases.

In discussing the question in these pages I have therefore tried to illustrate the influence of the antiseptic method on the healing of a wound, and on the prevention of septic diseases.

The aseptic, usually feverless, progress of a wound depends on the complete fulfilment of all the details associated with the method, and not alone on the prevention of decomposition.

In some cases treated antiseptically a rise of temperature takes place, or a simple aseptic wound-fever, so termed because it is not associated with decomposition within the wound. It is the kind of fever which occurs in cases of subcutaneous injury, it is altogether of a milder and more transitory type, and causes but trifling symptoms of general constitutional disturbance or none at all. It is probably caused by the absorption of the injured matter in the wound or injury; and the difference between this and septic forms of fever is that one is produced by the absorption of neutral substances, and the other by matter in a state of decomposition.

The second question is the degree of security afforded against the occurrence of wound diseases.

In respect of erysipelas, the antiseptic method appears to afford a relative, but not, I think, an absolute security.

Erysipelas occasionally occurs in connection with wounds in which no decomposition has taken place. We know that erysipelas will occur without any wound at all.

But the occurrence of erysipelas in connection with antiseptic treatment is rare.

That the method possesses any power in warding off tetanus we know as yet nothing.

Daily experience, more and more surely, establishes that in cases where the antiseptic method has been successfully carried out pyæmia and septicæmia are with certainty prevented.

Experiment has proved that carbolic acid is not so active as some other substances are in arresting the development of bacteria, yet the more powerful germicides have not proved on trial to be so advantageous in practice.

Carbol probably acts in some as yet unknown way on the albuminoid substances which are so prone to decomposition. It may so alter the chemical character of these substances that the presence of a moderate quantity of bacteria fails to excite decomposition.

Notably in some cases in which it fails entirely to prevent the presence of bacteria, it arrests their power of reproduction, and thus renders them harmless.

The Listerian method aims at perfection; but it cannot be in all cases absolutely perfect in the matter of air-exclusion. Some air must be included in the dressing itself, and the spray drives some germs against and into the wound.

The movements of the body, or the reduction in size of a part, favour the entrance of air beneath the dressing.

The volatility of the acid renders any but recently prepared gauze untrustworthy.

These are some of the elements of failure which interfere with the absolute security which might otherwise be obtained. However, a large experience has now proved that carbolic solution and spray, and the subsequent application of carbolised gauze dressings, do prevent decomposition in wounds and the consequences which arise from decomposition, better than any other method of dressing in use.

The antiseptic power of the material is longer maintained under Lister's dressing than in any other form; the duration of this will vary, it is curtailed by the occurrence of bleeding or a copious amount of discharge.

No kind of surgical injury is more common than fracture of a bone, and no distinction more clearly defined than that between simple and compound fracture. In the former, recovery usually takes place with facility. The injury in itself is by no means simple, the bone is broken, perhaps, into several sharp irregular fragments, the periosteum and muscles are torn, blood is extravasated into the tissues, the patient may have attempted to walk on his broken limb, and the soft parts are pulpified in consequence, yet so long as the skin remains unbroken, this very severe damage will not be followed by serious consequences. Adjust the fracture, immobilise the limb, and the pain, swelling, discomfort even, soon diminish and disappear, while the constitutional state differs little if at all from that of health. We do not much concern ourselves about the patient's age, or sex, or constitution; as a matter of course, all will go well. But without any other difference in the patient's condition, should there be a breach of surface, a wound communicating with the fractured bone, that which was before a simple injury, in so far as its consequences are concerned, at once becomes complicated by the possibility and frequent occurrence of grave peril to life or limb. Fever and suppuration commonly occur, and perhaps abscess, necrosis, erysipelas, septicæmia, or pyæmia.

As a rule, one fourth of the patients injured in this way used to die, others had to submit to amputation, the recovery of the rest was oftentimes most protracted, while some got well as easily and quickly as do cases of simple fracture. Generally those were cases where the external wound was small, and capable of being immediately closed. The great difference which exists between these two forms of injury, lies at the foundation of all wound treatment, and is the basis on which the antiseptic treatment of wounds has been erected.

For a long time, the action of the gases of the air was supposed to account for the difference; but we now know that pure air does not excite decomposition, but that there is a septogenic element in ordinary air, the precise nature of which, so far as practical results are concerned, is not material to the issue. Whatever it may be, it is something which can be separated from the air, leaving that medium incapable of exciting putrefactive change. The proof of this rests on some very simple and conclusive experiments.

The search began in enquiries into the nature of the poisonous material in the air, and the means of counteracting it; and in this a long series of experimenters have taken a part, amongst others Latour, Schultze, Schwann, Pasteur, Tyndall, Davaine, Koch, recently and notably Lister.

There can be no doubt that the air-dust does excite putrefactive changes in animal matter, for without changing air in any chemical way it may be rendered incapable of exciting putrefaction.

Schröder and Dusch established that it was not necessary to calcine air, or subject it to the action of strong reagents, like the mineral acids, that it could be rendered innoxious by simple filtration through cotton wool.

The experiments of Chevreuil and Pasteur showed that the deposition of this septogenic material took place by the mere action of gravity, and that air cleared of its organic dust might, for an indefinitely long time, remain harmlessly in contact with a highly putrescible fluid.

It has been long known that animal matter exposed under ordinary circumstances to the action of the atmosphere putrefies; but the discovery that it is not the atmospheric gases, but something suspended in them, which produces decomposition, is recent.

Pure air or foul air, so far as wounds are concerned, depends on the relative quantity, and probably the quality also, of the organic particles in the atmosphere.

It is the application of this fact to surgery which Lister made, viz. that not air itself, but something superadded, or suspended in it, of which it might be deprived, was the noxious influence acting on exposed or open wounds, and further as a corollary to this, he devised and finally perfected means for the purpose of destroying the septogenic elements already existing in a recent wound, and preventing their subsequent approach during the progress of the healing process.

The presence of micro-organisms in the atmosphere capable of multiplication *ad infinitum*, so soon as they find a suitable medium to grow in is certain; that organic matter generally, and wound-fluids in particular, offer a peculiarly fertile soil for their development, is also certain. It is further demonstrable, that where decomposition occurs there these organisms abound, but whether as cause or as effect is the still disputed question in the minds of many.

In a wound, however, decomposition becomes the starting point of a series of definite and dangerous diseases so distinctly connected with the change that they are called wound diseases. The antiseptic method aims at preventing the causes, and as a consequence abolishing the results, of wound putrefaction.

The decomposition of dead animal matter may be prevented in several ways. It will not putrefy if kept at a freezing temperature, if maintained at the boiling point of water, or if kept quite dry. The elements of putrefaction require for their development certain favourable conditions.

In all putrescent fluids we find bacteria, living organisms, with an indefinite power of reproduction. We do not know, however, if bacteria be the only agents capable of exciting putrefaction. There are apparently several different kinds; some potent for evil, others harmless, acting variously in different animals, and in different forms of disease.[1]

Whether they are themselves the essential agents, the mechanical poison, in short, or merely induce changes in the secretion or tissue, in consequence of some previous pathological change by which they are enabled to act, we can scarcely yet tell.

The germ theory of putrefaction assumes it to be capable of proof that putrefaction does not occur, except in the presence of these organisms, that it is produced by their agency, that they are always derived from similar organisms, and when implanted in a suitable soil, will multiply indefinitely.

Further, certain germicide agents absolutely arrest their power of reproduction, and with this arrest the so-called 'unhealthy action' in wounds either disappears if it have already occurred, or may be entirely prevented from taking place.

[1] '*Untersuchungen über die Aetiologie der Wund-Infections Krankheiten,*" Dr. Robert Koch, 1879.

This theory suits and explains most of the facts; and, until a better be forthcoming, we may accept it with a scientific reserve, but without hesitation.

Local inflammations, traumatic fever, the various forms of septic poisoning, depend therefore on wound-putrefaction. Certain substances, called antiseptics, hinder the development of micro-organisms, destroy the potency of those already existing, and arrest decomposition. There must be, therefore, first, the organism, or germ, or whatever it may be called, which we assume excites putrefaction, since it is invariably associated with it, and secondly, suitable conditions for its development. That putrefaction of the wound-secretions is produced by the introduction of micro-organisms, which gain access from without, the analogy of the action of the yeast plant in producing alcoholic fermentation, or the process of lactic acid fermentation of milk, goes a long way to prove. The micro-organisms either induce the change in animal substances which we call putrefaction, or are in some way most intimately associated with that change, since it is certain that when these micro-organisms are excluded, no alteration whatever occurs, while if they be allowed free access, the change under ordinary conditions invariably occurs.

Again, healthy, living tissue possesses certain powers of resistance. Experiment and clinical experience alike show that it will kill a certain proportion of septo-germs. Life, in fact, may be defined as a power capable of resisting decomposition. The persons of some men, and some conditions, and diseased states of the body, the diabetic for instance, are much more prone to admit putrefactive change than others. This it is which explains the comparative immunity of some persons, and some conditions, and accounts for recovery after certain forms of injury, or wound states.

A thin layer of lymph or blood between the surfaces of a recent wound, in a vascular, nervous part, does not necessarily decompose at all, but may become organised, as we constantly see in wounds of the face, and elsewhere; but the same thickness of lymph or blood, in contact with tissue that is not living, speedily putrefies.

The germ theory received rather a rude shock by the experiments of Klebs, Ranke, and others, which appeared to prove that organisms freely develop under antiseptic dressings, and were to be found even in healthy living tissues. Further experience and experiment tend, however, to negative these conclusions.

Cheyne [1] seems to have settled by very exact experiments that bacteria and micro cocci are different organisms, develop under different conditions, and are not interchangeable; that the former are intimately associated with the production of putrefactive change in animal tissue, while the micro-coccus is not; that these two forms of micro-organism thrive under different circumstances; and that in a cultivating fluid the addition of bacteria speedily renders it acrid, foul-smelling, irritating, and calculated to excite fever if absorbed into the system. The micro-coccus, on the other hand, changes the cultivating fluid very little, producing at most a faint, sour odour. When micro-cocci develop in wounds they occasion no smell in the discharge, no symptoms of local irritation, and none of general constitutional disturbance. Micro-cocci give little evidence of their presence, and are harmless, while the development of bacteria in a wound is followed by local and general disturbance.

Cheyne [2] never found bacteria under antiseptic dressings, though he discovered micro-cocci. But these organisms only appear at a late period, when the antiseptic dressing has lost much of its power; the acid being volatile, the dressing

[1] *Trans. Path. Soc.*, 1879. [2] Cheyne, *loc. cit.*

becomes too weak to resist their development, and they begin to appear, first at the margins, and thence in time spread further and further inwards.

In wounds not treated antiseptically, organisms, chiefly bacteria, are always present. Under the antiseptic treatment organisms of any kind are either completely absent from the wound, or, if present, exist only in the form of micro-cocci. A five-per-cent. carbolic solution at once arrests the movement of bacteria under the microscope; the septic properties of air-dust are destroyed by being soaked in the carbolic spray. Flasks containing putrescible fluids may be left open-mouthed in the spray for any length of time without result, while a momentary exposure to ordinary air afterwards will cause in these same flasks a speedy putrefaction of their contents. Micro-cocci develop more easily than bacteria in weak carbolic solutions.

These are some of the facts on which the principle of antiseptic surgery is based. All modern surgical practice is antiseptic in a degree; but the aim of the ordinary plans of treatment is rather to combat the consequences of putrefaction, than completely to arrest its development. As a rule, in open wounds some amount of decomposition occurs, and has hitherto been expected to occur. In a large proportion of cases a safe and successful result is accomplished by getting rid, as quickly as possible, of the products of decomposition, and preventing them from remaining in contact with the living tissue, which, to a greater or less extent, possesses a power of self-protection.

A perfect antiseptic method prevents putrefaction in wounds altogether; it tries so to act upon the tissues exposed in the wound as to place them in a condition as favourable for repair as if they had never been exposed at all, and it further furnishes during the period of that repair a protecting cover, which to some extent fulfils the office of the injured or absent skin; it is a method, in short, which aims at depriving a compound fracture of nearly all its peculiar risks, and converting it into one which behaves as nearly as possible as does a simple fracture.

Antiseptic protection cannot altogether prevent suppuration, but it excludes the septic air-dust, which is the most common cause of it. Suppuration may be excited by other causes; inflammation, arising in any way, is capable of producing it; irritation by reason of the antiseptic material itself may cause it, when it is applied too strong, or too directly and constantly to the wound-surface. Undue tension excites suppuration, as when imperfect drainage allows the secretions to accumulate within a wound-cavity.

The development of organisms in a wound may be more or less hindered in several ways. The wound may be purified frequently by the use of antiseptic lotions, the fluid in which the septo-germs grow may be allowed to flow away, or is washed away, as soon as formed, as in the 'open treatment' of wounds, or that by irrigation. The vital powers, general and local, may be increased, and at the same time the resisting power of the wound. The temperature of the inflamed tissues may be lowered by an ice-bag, or by the artificial anæmia produced by elevating the limb. In the most complete development of the antiseptic method the access in an active form of the putrefactive elements of the atmosphere is prevented during the continuance of an operation, or the dressing of a wound, by the employment of carbolic spray, or should they have already gained an entrance may be prevented from doing an injury by the employment of germicides. The wound having been thus rendered aseptic, may be so maintained, until it is healed beneath what is called an antiseptic occlusion; provision at the same time being made for the escape into the occluding antiseptic material of any secretions which may form. This covering must be of sufficient quantity, and renewed sufficiently often to neutralise any tendency to septic change existing in the discharge, and that such dressing changes when required may be safely accomplished, the wound, during the time of its exposure, must be kept enveloped by a carbolised atmosphere, sufficiently strong to neutralise the septic influence of the air.

Knowing that carbolic acid possesses powers for evil as well as good, we strive to limit the former, while not interfering with the latter. Formerly suppuration and wound-irritation were frequent consequences of the employment of carbolic acid in too strong a form. Now it is found that comparatively weak solutions suffice to prevent putrefaction, and produce little irritation.

The direct action of carbolic acid upon the diseased or injured surface is to be avoided so long as it remains aseptic. Some surgeons may still suppose they cannot have too much of a good thing, and daily syringe out the cavity of a wound treated antiseptically. The consequence is, they interfere with its repair. If the parts be rendered aseptic in the first instance, if proper drainage be employed, if there be no tension, then syringing is out of place, it can but wash away the layer of white corpuscles which is lying on the surface, preparatory to the repair of the injury, distend and irritate the cavity of the wound, cause hyperæmia and suppuration—the very things we are seeking to avoid.

At the outset there is little danger of using an excess of the antiseptic agent, but afterwards the direct contact of carbolic fluid should, as far as practicable, be avoided.

Wound-secretions offer an excellent cultivating fluid for bacteria, with whose advent in a wound appears decomposition, and with decomposition the possibility of the ill consequences which often follow it. Antiseptic materials arrest the development of bacteria, and the septic dangers of a wound. The spray used during the operation envelops each bacterium or germ, which falls upon the surface with an antiseptic coating, whilst it constantly washes away the animal fluids, and impregnates the superficial layers of the wound with the antiseptic agent, and thus renders the conditions most unfavourable for bacteric development. The less the wound-secretions are permitted to accumulate, the fewer are the chances of decomposition taking place. The farther bacteria are removed from living cells, the more active is their growth. If not too numerous, they lose their reproductive power and cease to exist in contact with living matter. In a clean-cut wound, whose surfaces can be kept in complete contact, any bacteria present will be killed by contact with living cells; but if a cavity remain, in which fluids accumulate, the stray organisms left have an opportunity of developing, hence the necessity for the close apposition of the wound surfaces, and of thoroughly good drainage, all chance of infection from without being guarded against by the use of means which, while hindering the development of organisms, cause as little injury as possible to the living tissues.

Wound-secretions of very high concentration, offer resistance to the development of bacteria. In this way healing under a scab may be explained. It is an aseptic form of healing: A scab which remains dry does not admit of the development of bacteria, and acts in a degree as a kind of antiseptic plug. Upon this fact the earlier efforts at antiseptic occlusion were founded. Compound fracture and other wounds with small external openings often heal in this way. The oozing of blood may have prevented the entrance of septic material, and the subsequent scabbing of the wound will often be followed by excellent results.

The long dispute as to the desirability of seeking primary union or otherwise, after an amputation, for example, must now of necessity come to an end. Those who left the wounds open did so because they observed that the edges might immediately unite, while the interior became filled with a fluid prone to decompose, causing much pain, inflammatory tension, bursting open the wound, perhaps, or becoming absorbed into the circulation, and producing symptoms of blood-poisoning.

Many expedients were devised to counteract these objectionable results. The wound was left open to 'glaze,' or not closed at all, until a protracted granulation process had healed it. No method is to be compared in efficiency with immediate closure and the employment of effective drainage, which allows the secretions to escape as they form, and facilitates, in many ways, the progress of repair.

It cannot be too plainly set forth that the antiseptic method is not the mere employment of any single reagent, such as carbolic acid, but the accomplishment of certain definite ends which demand for their fulfilment the avoidance of wound-putrefaction and its consequences, and the conversion of a dangerous open wound, to the comparatively safe and harmless condition of a subcutaneous injury. The simplest, best, and at the same time safest manner of obtaining these objects is the task before us. Carbolic acid has thus far proved itself well fitted for the purpose, and the now large experience of its use in the treatment of wounds proves that this ideal and most desirable result is one which may be consummated.

The mode in which a wound treated antiseptically heals may be well observed, if the dressing be daily changed for the purpose, in a wound of the scalp or other region accompanied with loss of skin and consequent exposure of the deeper parts.

In former methods of dressing, the wound had in the first place to 'clean,' and this cleaning was necessitated by the death of a superficial layer of tissue. The amount of 'cleaning' that was required, and the subsequent progress of the injury varied with the amount of violence inflicted, the condition and constitution of the patient, and the nature of his surroundings, and any tissue on the borderland between death and life was very likely to be involved in the process of molecular gangrene called 'cleaning.' Eight or ten days were required for the completion of the process, and by the time all the dead tissue was cast off, the granulations had been produced.

Something quite different is the rule under antiseptic dressings. In an incised wound, or one that is only moderately contused, and which has not sustained such violence as to cause the actual death of the tissue, this superficial necrosis does not occur, and the stage of cleaning does not as a rule exist. At first, there is more or less bloody serous and then clear discharge; after the fourth or fifth day the wound almost resembles a freshly made one. There is scarcely any secretion, no inflammation, redness, or swelling, and granulation and cicatricial tissue are produced without suppuration. The same process will occur to some extent even in exposed bone, in which there will be less likelihood of necrosis, and if it do occur it will be less extensive.

Before the introduction of the antiseptic method, if a wound did not heal by the first intention, or could not be united by suture, it suppurated, as a matter of course, in about three or four days. This does not now occur, the gap is gradually filled up with new material, which becomes organised into cicatrix, and from first to last there may not be one drop of pus.

Sometimes we see a drainage tube filled with a greyish substance, or the surface of a wound coated with it. This is what is called plastic lymph; in other circumstances, pus might have been formed instead. This pultaceous matter, however, if not disturbed, soon becomes vascular, and will often fill a considerable wound-cavity.

Again, a blood-clot filling the hollow of a wound does not soften down nor become septic. It may for an indefinite period remain unchanged in appearance on its surface, which is dark; but in its deeper portions it becomes gradually transformed into connective tissue and cicatrix, by the multiplication of the white cells in it and the immigration of others.

If the injury be such as to kill the implicated part, or if the margin of a flap or a piece of tissue which it has been attempted to save become gangrenous, the amount of necrosis which takes place in the part and the changes set up for the purpose of getting rid of the dead tissue will be strictly limited, the dead portion during its separation will not be a source of danger to the adjacent tissue, nor to the individual; it will excite neither inflammation nor suppuration, but will probably shrink or dry up, and be in time thrown off with the least possible disturbance, or if not too extensive, may even become absorbed and never separate at all.

In a wound whose edges cannot be brought together, or which have separated from some accidental cause, healing may still take place without suppuration, either by the gradual organisation of a blood-clot or by the production of plastic matter, whose elements are much the same as those of the blood-clot *minus* the red corpuscles, which do not, however, take any part in the healing process. From this will result, it may be, cicatrix of considerable breadth. It is quite certain such a transformation, and such a result do not occur under the ordinary methods of wound-dressing.

ON ANTISEPTIC MATERIAL.

THE materials in most common use are, carbolic acid, salicylic acid, boracic acid, thymol, chloride of zinc, acetate of alumina.

A perfect antiseptic material is yet a desideratum; carbolic acid is, so far, the best we have, but it irritates, and occasionally produces poisonous symptoms, which, in some instances have terminated fatally, and in other cases lead to the necessity for abandoning the carbolic method of dressing. The weaker the strength used—consistently with obtaining the required result—the better for the wound, and the better also for the patient. A non-poisonous, non-irritating antiseptic, possessing the advantages otherwise of carbolic acid, is what we hope to obtain at a future time.

CARBOLIC ACID, OR PHENOL, C_6H_6O.

This was the antiseptic first employed by Mr. Lister, and it has remained in general use ever since. It is a product of the destructive distillation of coal, and can be made in other ways. It is really an alcohol of the phenyl series, and may be regarded as the hydrate of phenyl, just as common alcohol is the hydrate of ethyl. Absolute phenol is simply pure carbolic acid. It is more readily soluble in water than the ordinary commercial acid, which contains an homologous substance, called cresol. Cresol does not crystallise, and, though very deliquescent, does not dissolve readily in water; it possesses antiseptic properties similar to phenol, but is more irritating, and causes anæsthesia, and tingling of the skin of the hands, in a much greater degree than the pure acid does. Most of the commercial gauze is prepared with the less pure form. Absolute phenol is a fine, crystalline powder, readily soluble in fifteen parts of water at the ordinary temperature. Its less irritating properties are a great advantage, while the odour is less disagreeable, and more evanescent, but it costs considerably more than the ordinary acid.

Carbolic acid is employed in different ways. In watery solutions of various strengths—mixed with oil or spirit—in the form of spray, and of antiseptic gauze. It is often convenient to preserve it for use in a fluid form. If 500 grammes of the crystallised acid be heated with 28·3 grammes of water, it will remain fluid, and each cubic centimètre will contain one gramme of acid, so that the quantity required may be measured, in place of being weighed.

Unfortunately the decimal system does not apply to our more complicated system of weights and measures.

The watery solutions are of two strengths: one being 5 per cent. or 1 part in 20, and the other 2½ per cent., or 1 part in 40. One part of the ordinary commercial acid will dissolve with some difficulty in 20 parts of water.

The five-per-cent. solution is employed for purifying the hands of the surgeon and his assistants, before, and during an operation ; also to disinfect the surface of that region of the body where the operation is about to be performed, and all parts which will be included in the subsequent dressing —also for supplying the steam spray. In a solution of this strength, sponges are preserved, also silk, and drainage tubes. Catheters are cleansed, and may afterwards be kept in it.

The 1 in 40 solution is used for the purpose of irrigating a wound, washing the sponges used during an operation, soaking the gauze which is first applied to the surface, and for filling the tray in which the instruments required are placed. Glycerine in equal proportion to the carbolic acid may often be added with advantage to the watery solution. It helps to prevent the too rapid volatilisation of the acid, and counteracts, to some extent, its irritating properties.

Water possesses no very strong attraction for carbolic acid; the latter is readily given off by it, hence watery solutions seem to act more intensely on the skin, or any surface to which they are applied. When the acid does not dissolve in 20 parts of water, but partly remains suspended in the form of oil-globules, impurity to that extent is indicated, and the solution should be filtered before using, as the undissolved particles act as a caustic.

The solution intended for spray purposes should always be filtered.

CARBOLIC OIL.

Carbolic acid readily blends with oil, in any proportion. The oily mixtures are two in number, and are preferably prepared with olive oil, the weaker being 1 in 20, or 5 per cent., and the stronger 1 in 10, or 10 per cent.; the former is used for oiling catheters, specula, hands, and fingers, while in the stronger mixture, lint may be soaked previous to introduction into a deep wound, such as is produced in a necrosis operation.

The best way of preparing either oily or watery mixtures is, to first put a few ounces of oil or water into the jar or bottle, and then add the full quantity of acid, previously melted by heat. Mix the two thoroughly, and afterwards add the remainder of the oil or water, otherwise it is difficult to properly blend the oil or water with the acid.

Carbolic acid dissolves with great facility in alcohol. A

strong spirit solution, 1 in 5, caustic in character, is sometimes used as an application to wounds which have been exposed for a time to septic influences. A more useful spirituous solution is 1 in 10 or 10 per cent. in strength. It is a very powerful and efficient antiseptic, and has been used in injuries of the head, and elsewhere.

CARBOLISED GAUZE.

One of the most important ways of employing carbolic acid is, suspended in gauze, by means of a mixture of paraffine and resin. This material may be bleached or unbleached 'book-muslin;' it is conveniently prepared in lengths of six yards, which, when thoroughly impregnated with the mixture containing the acid, are subjected to strong compression at a high temperature to diffuse the mixture equally, and express any superfluity, and then having been dried in a warm room, they are preserved in air-tight cases for use. It is an advantage, but adds considerably to the cost, if gauze be used from which the fatty particles have been thoroughly removed by the action of a caustic lye. It will then not only absorb the antiseptic mixture much more readily, but also the discharges from a wound. Bleached gauze is a nice material to look at, softer and more pliant but also more expensive than the ordinary kind.

Professor Lister recommends that the gauze should now be prepared by saturating it with a mixture consisting of:

	Parts.
Carbolic Acid	1
Resin	4
Paraffine	4

This contains half as much more carbolic as the former gauze, and, while it is stronger as an antiseptic, is not more irritating. The amount of paraffine is less, and the gauze is therefore a little more sticky, which tends to keep the dressings more firmly in place. The resin is used because of its strong retentive power; it arrests the rapid volatilisation of the carbolic acid.

The paraffine prevents the layers of gauze from adhering together, or to the skin. It should be pure, as crude paraffine acts upon the india-rubber in the mackintosh, which it renders soft and useless, and has also a tendency to produce the eczema of the skin usually attributed to carbolic irritation.

A considerable experience in the use of the antiseptic gauze affords proof, that to be thoroughly trustworthy, it must be freshly prepared; that however carefully kept, if it be for a long period, the greater part of the carbolic acid will escape. One cannot doubt that many of the failures attributed to the antiseptic method have been really due to a deficiency of the antiseptic reagent in the material employed. Freshly prepared gauze is supposed to contain from 6 to 9 per cent. of carbolic acid; but however carefully it be kept, it loses a large portion of this in a month or two.

According to Münnich's experiments, gauze or jute containing six or seven per cent. of carbolic acid when first applied, contains after two days' application as a dressing from 2 to 3 per cent., after four days' 1 to 2 per cent., and after seven days' still less. Amongst the various kinds of gauze experimented on, that prepared with paraffine possessed the greatest retentive power, and showed no signs of chemical change.

Generally speaking, the gauze is always purchased ready made; but for those who wish to manufacture it for themselves, several good formulæ are available.

The modifications of Bruns and Küster offer facilities, as they do not require any complicated apparatus, nor the pressing and long period to dry in a special chamber, which the English gauze does. The following are the proportions used in the preparation

	Parts.
Rectified Spirit	1000
Resin	100
Carbolic Acid	100
Glycerine or Castor Oil	100

The gauze may be saturated with the mixture by means

FIG. 1. Küster's apparatus for the ready preparation of carbolic gauze. It is of wood, and can be made by any carpenter. The gauze as it is unrolled from one side passes slowly through the fluid beneath two rollers which keep it immersed, and is then slowly wound upon the reel on the opposite side.

The weight which is slung over the roller prevents the gauze unwinding too fast; while, on the opposite side, a flange stretches the gauze as it passes up to the reel, and removes any superfluity of the carbolic mixture, which is thrown back into the trough by a slab projecting beneath the reel.

of a simple apparatus such as that shown in the drawing, and after being hung up to dry for a period of six hours,[1] is packed ready for use in an air-tight case. This gauze is

[1] If needful, a much shorter period would, no doubt, suffice.

very pliant, not irritating, it is cheaply produced, and contains a very large proportion of carbolic acid. There are however, some disadvantages in the use of both castor oil and glycerine. The former is apt to partly decompose into fatty acids during warm weather; while the hygroscopic character of glycerine permits the somewhat too ready absorption of fluids, which pass directly to the surface of the dressing, in place of spreading uniformly throughout the layers. In practice, however, these objections have not been strong enough to interfere with the successful employment of the material so prepared.

A very excellent soft gauze, possessing none of these drawbacks, may be made as follows:

	Kilogrammes.	Mètres.
Gauze	1	= 37
		Grammes.
Resin		400
Stearine or Paraffine		60
Glycerine		80
Spirit		1200
Carbolic Acid		100

To increase the antiseptic power of this mixture 100 grammes of boracic acid may be added, (it does not crystallise out in this proportion). Müunich has found that a substantial, sufficiently fine meshed gauze does not require the troublesome and expensive process frequently employed to remove its fatty particles; if future experience bear out this opinion, the gauze will be more cheaply and easily prepared.

Bruns has devoted much time to the discovery of a good formula for the production of cheap and efficient gauze. His last is:

	Grammes.
Carbolic Acid	100
Resin	400
Castor Oil	80
Spirit	2000

400 grammes of powdered colophonium or resin are slowly added, with constant stirring, to 2,000 grammes of spirit.

This dissolves in 10 to 20 minutes, and when solution is complete, 100 grammes of carbolic acid and 80 grammes of castor oil are added, stirring being continued. In place of the castor oil, 100 grammes of melted stearine, or 100 grammes of glycerine, may be used.

The quantity of mixture thus prepared suffices for the impregnation of 1 kilogramme of gauze (37 mètres). This is placed, folded in a flat vessel, the mixture being poured over it, it is then kneaded until thoroughly soaked. It is afterwards hung up, to allow the evaporation of the spirit, and is dry enough to be ready for use or storing in from 5 to 10 minutes, according to the temperature of the air. The whole operation after the mixing of the solution may be completed in from 15 to 20 minutes. The gauze contains about 9 per cent. of carbolic acid.

Bruns has suggested that a concentrated solution, easy of carriage, from its small bulk, might be utilised in military surgery, the needful amount of spirit for dilution being added just before using.

The formula is:

	Parts.
Carbolic Acid	25
Resin	60
Stearine	15

This forms a mass of soft salve-like consistence, 80 parts of which are readily miscible with 100 of spirit. A comparatively small quantity, 675 grammes, added to 2 litres of spirit, suffices to prepare in a short time 25 mètres of gauze containing a large proportion of carbolic acid.

Another formula is:

	Parts.
Carbolic Acid	100
Resin	400
Spirit	100
Ol. Ricini, or	80
Melted Stearine	100

The resin, very finely powdered, is added gradually with constant stirring to the alcohol, and then the carbolic acid, and castor oil or melted stearine are introduced. This forms a semi-fluid mixture, of the consistence of honey, which keeps well, and dissolves readily in spirit. When required for use, if made with castor oil 680 grammes, if with stearine 700 grammes are added to 2 litres of spirit. The mixture so obtained suffices to impregnate 1 kilo. of gauze in the way already described.

Carbolised jute can be similarly prepared, and is considerably cheaper, but it is less convenient; and with regard to its employment in military surgery, Bruns points out that gauze takes up less room; and that the quantity required for dressing is much lighter. Thus for an amputation of thigh, 200 grammes of jute are required, but only 82 grammes (2 square mètres) of gauze.

When the gauze is preserved in a strongly compressed form, it retains the carbolic acid for a longer period. After three months Bruns states that as much as 8 per cent. of carbolic acid has been found in it.

CARBOLIC ACID POISONING.

This may occur in two forms, either acute or chronic. In the severe form it is rare, at all events in this country, but the minor degrees are not uncommon.

Some persons seem peculiarly susceptible to the influence of carbolic acid, and in them quite a small quantity will suffice to excite symptoms of poisoning. Children and women seem more especially liable to the noxious influence. It may occur from the mere inhalation of carbolic acid, but it is much more frequently observed as the result of washing out large cavities or wounds. Care must be taken never to inject carbolic acid forcibly, especially where the region

CARBOLIC ACID POISONING.

abounds in loose connective-tissue, the risk of poisoning being thereby greatly increased.

The impure forms of the acid would also appear to possess greater toxic power.

In severe or fatal cases, there are formidable collapse, great depression of temperature, pallor, thready, extremely rapid pulse, and feeble respiration, by the failure of which last, death appears to take place—the heart's action continuing for some time. The temperature falls until the time of death. Salivation may occur; mydriasis also is often present.

In the less severe and chronic types the symptoms are not very characteristic: gastric derangement, nausea, vomiting, and loss of appetite, together with more or less fever, are common symptoms, also giddiness, visual disturbances, noises in the head, more or less stupor. There may be some degree of collapse, with cold pale surface, feeble pulse and exhaustion.

Carbolic acid is quickly eliminated by the kidneys, and appears to excite irritation in them to such an extent as in some cases to produce albuminuria.

From this it would appear that the employment of carbolic acid may prove dangerous or mischievous in persons affected with chronic kidney-disease.

The quantity of urine is diminished in carbolic acid poisoning, and very often changed in colour. Vesical catarrh in varying degree may also be induced. The dark-coloured urine indicating carbolism may be passed in this condition, or only become dark after standing for some time. The appearance is probably due to the colouring matter in the blood escaping, in the shape of an increased amount of indican. There is no direct relation between the toxic effects of the carbolic acid and the amount of the discoloration. It is met with where there are no other symptoms; or the urine may be clear, while other well-marked acid signs of carbolic poisoning are present.

The sulphates are absent from the urine. In doubtful cases of carbolism when the discolouration of the urine is absent, the chloride of barium test will indicate this deficiency, and perhaps enable one to discriminate the collapse of carbolic acid poisoning from that due to the effects of the anæsthetic or the loss of blood.[1]

The general symptoms may occur immediately, or after an interval of some hours. There is no treatment known to be of any service. When severe the dressings must be at once changed. With discontinuation of carbolic acid, time is probably the best remedy. Heat should be applied to the extremities, and subcutaneous injections of ether, when there is collapse. Sulphate of soda or glauber's salt, has been recommended as an antidote internally, but it is of doubtful benefit. Some severe or fatal cases of so-called shock occurring shortly after an operation, may in part depend on carbolic acid poisoning.

Carbolic eczema and erythema sometimes result from long-continued application of carbolic acid to the skin, and the crude paraffine often used in the manufacture of the gauze will also cause it. Some persons are much more susceptible, more thin-skinned, than others. If at all considerable in amount, it is frequently associated with a certain amount of fever. A rise of temperature, in cases treated antiseptically, may sometimes be accounted for in this way.

---

[1] Vulpius, in the *Pharm. Zeitung* March 5, 1879, says that in cases of carbolic acid poisoning the urine ceases to contain sulphates. He therefore proposes as a test the addition of barium chloride to the urine previously strongly acidulated with HNO_3, when, the characteristic insoluble precipitate of barium sulphate will be absent.

If urine containing carbolic acid be poured on sulphuric acid in a test tube, a dark brown layer will form between the acid and the urine.

JUTE.

Jute is very largely used abroad as a raw material in place of the more expensive gauze. It is the fibre of certain species of Corchorus, chiefly the *Corchorus capsularis*, which is used for spinning and weaving coarse textures. The fibres are whitish, and consist of from 30 to 100 fibrillæ, flattened in external shape, and tubular in the centre. They are for this reason well adapted for the absorption and retention of the carbolic medium, which penetrates both between the fibrillæ and within them. The jute requires to be cleaned before impregnation. It is conveniently prepared in pads of five, ten, and fifteen grammes each; these are soaked for an hour at least before using in a five-per-cent. solution.

Prof. Bardeleben, in the Charité Hospital, Berlin, has been in the habit of using moist carbolic dressings, composed of jute or tow, which have been first soaked as described in five-per-cent. carbolic, and then preserved for use in a 1·40 solution; before application they are wrung out, comparatively dry, and a wet, carbolised bandage is employed to fix them. They require much more frequent changing, however, than gauze dressing does, irritate the skin a good deal, and are liable to produce symptoms of carbolic acid poisoning.

Münnich has recently, both on the score of cheapness and the objectionable character of the wet dressing, attempted to produce a dry jute dressing. The impregnating mixture is made up of—

	Parts.
Carbolic Acid	50
Resin . .	200
Glycerine .	250
Spirit . .	550

One-pound portions of tow are dipped into the mixture,

and afterwards pressed and dried. They contain after drying eight per cent. of carbolic acid, which is retained if they are well packed in parchment paper. Specimens tried lost two per cent. in three months, and four and a half after six months' keeping.

The absorbent power of tow, however, is scarcely good enough to take up the wound-secretions perfectly, and the application requires frequent changing.

Münnich finds that jute dipped in a ten-per-cent. solution of carbolic acid in spirit dries quickly and retains a large quantity of carbolic acid, as much as six to eight per cent. for ten or twelve days.

SALICYLIC ACID, $C_7H_6O_3$.

This agent was introduced by Prof. Thiersch. It can be made by heating carbolic acid together with caustic soda, and passing a stream of carbonic acid through the liquid. By this process salicylate of soda is formed, from which salt the salicylic acid is readily set free by the addition of a stronger acid, which combines with the soda. It was formerly obtained by the action of caustic potash on the bitter principle of willow-bark salicin. It is a white, odourless, feathery, crystalline substance, which requires 300 parts of cold water for solution. Boiling water will dissolve 4 per cent., and alcohol takes it up in large quantity. It is an antiseptic of less power than carbolic acid, but has the advantage of not being poisonous. Wounded surfaces are not irritated by it, nor the granulating process interfered with. It is employed in watery solution, or in the form of salicylic wool, which is usually made of two strengths, 4 per cent. and 10 per cent. To prepare this wool, all its fatty matter must first be removed, which makes it very absorbent of any fluid or of discharges. A piece of such wool thrown on the surface of water rapidly sinks, while a piece of ordinary cotton wool will float upon the surface indefinitely.

The stronger quality is thus prepared, 10 kilogrammes of the wool suffice for one kilogramme of salicylic acid, dissolved in 10,000 grammes of spirit, mixed with 60 litres of water. It is convenient to saturate 3 kilos. of the wool at a time in this mixture, which is placed for the purpose in flat wooden vessels. In these the wool is laid in single layers, each being thoroughly saturated before the next is superimposed. When the whole mass has thus soaked for about ten minutes, it is turned upside-down, the layers taken off in the order that they were put on, and laid aside to dry, flat, and in a warm room. The acid is deposited in the interstices of the wool, in the form of fine crystals. It is difficult to diffuse a quite uniform quantity of the acid throughout, and on drying, the crystals are liable to shake out. This may be remedied, however, by adding a small proportion of glycerine. The amount of salicylic acid in any specimen of wool may be estimated by its capability of neutralising a volumetric solution of caustic potash or soda. The commercial salicylic wool is a very uncertain preparation. Wiebel, chemist in the State Laboratory at Hamburgh, has shown that it often contains only one-fourth of the designated quantity of acid. He considers that this is not so much the result of shaking out of the crystals, as the use of warm-water solutions for the impregnation of the wool: as much as half the acid being sometimes driven off by the heat employed.

The wool may be used as an antiseptic dressing by itself, in cases where carbolic acid is not applicable, or in combination with carbolic gauze. When used alone, a thin layer of wool, dipped in ten-per-cent. salicylic glycerine, is first applied to the wound, with a piece of protective interposed; then a thick layer of wool, extending widely beyond, on all sides, is applied and the whole covered with mackintosh.

The most useful application, however, of salicylic wool is, as a supplement to carbolic gauze, when it may be used to fill up all irregularities, and to interpose at the margins of the dressings; it thus forms an efficient barrier to the admission of air, and is a further safeguard against septic influences, especially in places where there is much body-movement.

For cheapness, Thiersch has prepared a salicylic jute. He first dissolves 64 grammes of salicylic acid in 32 grammes of spirit, with the aid of gentle heat; the solution is then mixed with 128 grammes of glycerine, and the mixture thrown into a litre of hot water. This suffices to prepare half a kilo. of jute. The jute is placed in the fluid, which it almost entirely takes up, and is then hung up to dry; any fluid draining out is subsequently again poured over the jute, so that the whole of the salicylic acid is eventually taken up; the material contains about 3 per cent. It remains slightly moist because of the glycerine and the crystals of the acid do not so readily fall out. The disinfection of the surface, the hands, sponges, &c., is effected with salicylic lotion; but the instruments must be placed in carbolic lotion, as salicylic acid oxydises the steel. Salicylic solution answers very well to wash out large cavities. I have used it with much advantage, mixed with vaseline, 1 in 10, as an application to burns or granulating surfaces, and for smearing catheters or bougies.

Kolbe states, in proof of salicylic acid being absorbed, that its presence may be detected in the urine by the addition of chloride of iron, which gives a violet reaction. It is also said, when used over large surfaces, to impart a greenish tinge to the urine, but it never produces poisonous symptoms.

THYMOL.

Thymol is also one of the phenyl series. It is a camphor-like substance, very volatile at ordinary temperature; it mixes easily with oil or alcohol, but 1,000 parts of water are required to dissolve it. It is a less powerful antiseptic than carbolic acid, and on further trial has not borne out the strong praise given to it by Ranke.[1]

Its chief advantage, save that it is not poisonous, is that it does not interfere with the progress of cicatrisation. It produces much smarting pain after being applied, either in the form of spray or lotion; it is not strong enough to disinfect a wound, a granulating surface, or the hands and instruments. In the form of gauze, which can be prepared in a similar fashion to carbolic gauze, it may be used safely for small operations, or where very little discharge is expected. I have failed to discover that it possesses any advantage over carbolic gauze; while it is certainly far from being so safe or efficient as an antiseptic. It is also costly.

BORACIC ACID, H_3BO_3.

Boracic acid is an excellent antiseptic application to superficial granulating surfaces. It is found in volcanic regions in various parts of the world, and in borax or borate of soda, a substance used by the ancient Arabians. Boracic acid is a white crystalline, non-volatile substance, quite bland, unirritating, and not poisonous. It may be used in the form of boracic lint, boracic lotion, and boracic oil, or ointment. A proportion of 3·63, or nearly 4 per cent. of the acid saturates water, and forms a boracic lotion, which being quite clear should be tinted with some colouring matter, as the

[1] '*Ueber das Thymol und seiner benutzung bei der antiseptischen behandlung der Wunden*, Von Hans Ranke, Sammlung Klinischer Vorträge, 1878.'

tincture of litmus, which gives it a purple tinge (ʒj to Oj.). A much larger quantity of the acid dissolves in warm water or alcohol. Boracic lint is prepared by soaking the lint in a strong, nearly boiling solution, containing about 30 per cent. of the acid, and then allowing it to dry. It should contain nearly its own weight of the boracic acid crystals. It is also tinted, to distinguish it from ordinary lint. It forms a good application for any superficial and granulating sore, or for ulcers or burns. To these it may be applied after moistening in boracic lotion, protective being first interposed. The lint should freely overlap the exposed surface in all directions. It is sometimes prepared of a weaker strength by dipping in a cold saturated solution and drying.

It is an excellent dressing in cases of skin-grafting: which may be very successfully performed by first purifying the surface from which the grafts are taken, by 1 in 20 carbolic solution, then after the small pieces of skin have been divided and placed on the surface to be healed, securing each graft in its place with a narrow strip of protective, and covering the whole surface with boracic lint.

Boracic lotion may be used in the form of spray in children; it is a good injection in gonorrhœa, and as a cleansing lotion for the mouth.

Boracic ointment is formed with boracic acid and white wax, 10 parts each; almond oil and paraffine, 20 parts each. Melt the wax and paraffine, in the oil by heat, then stir in the finely-powdered acid until the mixture cools.

Boracic acid may be usefully blended with vaseline in 5 and 10 per cent. strength. Either of these forms a bland, unirritating application, which protects the surface of healing wounds, can be easily removed, and allows the secretions to flow away readily beneath it.

ACETATE OF ALUMINA, $Al_2O_3, 2C_4H_6O_3, 4H_2O$.

Acetate of alumina is a white amorphous deliquescent substance. It is an excellent antiseptic, more powerful even it is alleged, than carbolic acid; but it can only be used in the moist form. It is cheap, unirritating, and non-poisonous.

The hydrated alumina may be economically prepared by precipitation from a solution of common alum by carbonate of soda; and this can afterwards be dissolved in acetic acid.

The commercial method of obtaining the acetate of alumina is, to mix acetate of lead in solution with alum; the sulphate of lead precipitates, while acetate of alumina remains in solution.

Loewig has recently formulated a method by which a solution of the constant strength of 15 per cent. may be prepared. 10 parts of hydrate of alumina are mixed with 8 parts of dilute acetic acid, and allowed to stand for 24 to 36 hours, at a temperature of 30° to 40° R. (42°—50° C.) The filtered solution thus obtained is of fifteen-per-cent. strength.

Prof. Maas of Freiburg has recently largely used this material in a great variety of cases, and with very satisfactory results. Maas uses a spray of $2\frac{1}{2}$ per cent.; the compresses are dipped in a solution of similar strength, the wound being covered with protective. The dressing is completed by the application of compresses in sufficient number and thickness, and the exterior is covered with mackintosh. The amount of wound-secretion is usually very small, as the substance is almost neutral; the progress of Maas' cases was thoroughly aseptic, and very few dressings were needed. In a dry state the acetate of alumina decomposes, acetic acid being given off.

CHLORIDE OF ZINC, $ZnCl_2$.

Chloride of zinc is a white, very deliquescent substance, very soluble in water. It forms the active ingredient in Burnett's Disinfecting Fluid and Canquoin's Caustic Paste, and is a very powerful antiseptic. It is commonly used in a strength of 8 per cent., or 40 grains to the ounce. The solution appears milky, from partial decomposition of the zinc, when the solution will not be of full strength. It is employed as an application to wound-surfaces, especially when there is any suspicion of septic change. Also after the removal of malignant growths, for which purpose it was introduced by the late Mr. Campbell de Morgan. A superficial layer of tissue is cauterised by the application, and the white translucent film produced by the coagulation of the albumen forms a perfectly unabsorbent surface, capable of protecting the parts underneath for some time.

After long-continued use of carbolic dressings, the surface of a granulating wound becomes unusually absorbent; and when the carbolic acid is left off, a certain degree of fever frequently follows, probably due to some absorption of matter from the wound. This capacity for absorption continues for two or three days after the disuse of the carbolic. Prof. Maas and Dr. Hack [1] have recently published investigations as to the capacity for absorption of various substances, possessed by a healthy granulating surface; and ample proof is presented by them that a layer of granulation tissue must afford a very insecure protection against the absorption of wound-fluids. The chloride-of-zinc solution may be used in cases where it is necessary to afford a protection of this kind, and it furnishes an absolutely secure one—the effect

---

[1] Professor Maas and Dr. Hack, '*Deutsche Zeitschrift für Chirurgie,*' vol. xii. 1879.

persisting for two or three days. It is also probably the best and most powerful antiseptic we can employ to disinfect wounds which have for some time been exposed to septic influences, or in which septic change has already taken place. For this purpose it may be applied with advantage in solution or in the form of chloride of zinc jute or lint, which may be prepared in the following manner :—

Ten parts of chloride of zinc are dissolved in 100 parts of water, without admixture of hydrochloric acid, as this acts injuriously on the vegetable fibre. This quantity will suffice to soak 100 parts of jute, or lint, which should be well kneaded in the fluid, and then dried. It dries perfectly, in from 36 to 48 hours although the salt itself is so very deliquescent. It may be preserved for any length of time without change; the chloride of zinc does not crystallise in the medium, nor is it likely to fall out, or be irregularly distributed, like salicylic acid. It has been used by Bardeleben, as a useful dressing even in operation-wounds, on which the caustic effects of the zinc appear to be very slight. Chloride of zinc may be used to disinfect sinuses, or as an application to wounds near the orifices of the mouth or anus, when its valuable properties may tide over the immediate dangers of septic absorption, and protect the surface at the time it is most exposed to risk from toxic influences.

PROTECTIVE.

The object of this substance is implied by its name; it prevents the immediate contact of the carbolic acid with the wound, neutralises its irritating properties, and should be itself perfectly neutral and unirritating.

It should consist of oiled silk, coated on both sides with a thin layer of copal varnish, which renders the material impermeable to carbolic acid vapour. The surface of the protective is further covered with a layer of dextrine, by brushing it over with a solution containing 1 part of dextrine and 2 of starch, dissolved in 15 parts of five-percent. carbolic solution ; when this dries, it forms a coating over the otherwise repellent surface of the varnished silk which will now readily and uniformly take up the 1·40 carbolic solution, in which the protective is always first dipped, before applying it to the wound-surface.

At the temperature of the body the material becomes quite soft and pliant, moulding itself closely to the surface. The protective should cover, and slightly overlap the margins of the wound, being in turn overlapped by the layers of folded moistened gauze. The protection of the wounded surface from the direct action of the carbolic acid is very important, since this substance has of itself a tendency to irritate the wound, to promote the occurrence of suppuration, and to hinder repair.

The silk protective contains a trace of lead, and this causes it to become dark in colour, stained in fact by any discharge which is decomposed. In a wound which continues aseptic, the protective should remain without discoloration, unless influenced by the colouring matter of blood, or by the sulphur in the india-rubber drainage tubes, which may sometimes occur. If sufficient pains be taken to disinfect it, the same piece of silk may be used several times.

MACKINTOSH.

This is a thin, impermeable material, formed of cotton coated with india-rubber. Its purposes are, to arrest the evaporation of the volatile carbolic, to prevent the secretion which soaks through the dressing, if it should come directly to the external surface, from being there exposed to atmospheric influence, and becoming septic, in close proximity to the wound. The air-proof quality of the mackintosh prevents this, until the secretions have reached its margins, and thus have travelled its entire width, which needs a considerable time, and also that the discharge should be considerable in quantity. The appearance of discharge at the margin of the mackintosh is the ordinary test of the time for changing a dressing. As a rule, a dressing should always be changed before it has thus soaked through. For convenience of application, the mackintosh after it has been cut the proper size, is placed beneath the outermost layer of dry gauze, the other eight layers, of which it usually consists, intervene between it and the moist gauze applied immediately over the protective. Gutta-percha tissue, oiled-paper, and other substances have been used as substitutes, but they are not nearly so efficient. The same piece of mackintosh, if thoroughly cleaned and disinfected, may be used several times.

Care should be taken to see that the mackintosh is sound, without a rent, pin-hole, or other defect. The presence of a small opening more or less completely neutralises the object, and may entail entire failure of the antiseptic dressing.

BANDAGES.

Bandages made of the carbolic gauze are cheap, light, adjust themselves perfectly to the part, do not readily slip, as each turn adheres to the previous one, and may be applied either moist or dry.

They are usually six yards long, and of various widths. They should always be dipped in 1·40 carbolic lotion, when applied to retain the deeper parts of the dressing *in situ*: for it may be taken for granted that the gauze bandages from their method of preparation and the necessary exposure contain little, if any, carbolic acid, while dust or septic matter finds a ready lodgment in the meshes of the material.

ELASTIC BANDAGES.

These are employed to fasten down the margins of the dressing in places where the movements of the body tend to loosen the dressing, and thus allow air to penetrate beneath its margins. These bands are made of ordinary elastic webbing, they are most manageable in short lengths, and should vary in width, from 1 to 2½ inches, according to the size of the dressing, or the region of the body to be dressed. They should not be put on too tightly, but just stretched firmly enough to keep everything secure against the movement of the part, or the shifting position of the patient. In dressings applied to the trunk, the axilla, neck, head, or the groins, their employment is an absolute necessity, and illustrations of the mode of their application are given in the chapter on Antiseptic Practice.

ASEPTIC SUTURES AND LIGATURES.

These may be prepared of carbolised catgut, silk, and horsehair, or silver wire may be used in many cases with advantage.

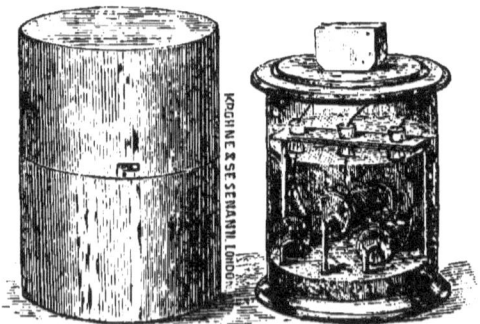

FIG. 5. Catgut Bottle.

The gut is prepared from the small intestine of sheep, and is preserved for use in five-per-cent. carbolic oil). A bottle such as that in fig. 5, with a case in which it can be safely carried about, will contain two or three sizes of gut, and is perhaps the most convenient and certainly the surest way of keeping it. A catgut holder, shown in fig. 6, may be

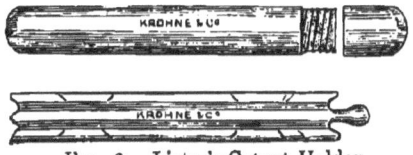

Fig. 6. Lister's Catgut Holder.

kept in a pocket-case. For daily use in hospital, reels holding the gut may be kept in a trough of carbolic oil covered with a lid like that shown in fig. 7.

The catgut is often carelessly prepared, and not thoroughly aseptic. Several months' steeping, at least, are required. Good sound thread should be chosen, and steeped in the following emulsion:

	Parts.
Pure Phenol	10
Water	1
Olive Oil	50

First, melt the phenol in the water with the aid of

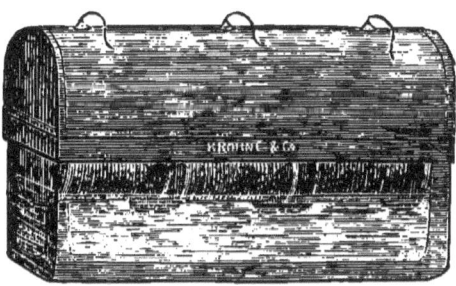

Fig. 7. Glass trough, containing carbolic oil and three reels of gut, with a perforated lid for the thread to be drawn through.

some heat, and then add the oil, shaking it thoroughly. The addition of water is necessary, as without it, the gut remains soft and will not tie a secure knot. In its ordinary state, gut will not tie a fast knot, it is slippery and swells when moistened. But the long-continued steeping in the emulsion altogether changes its character, it becomes firm and translucent. Steeping for five or six months is needed to transform the gut into good suture material, which does not slip when tied, and is quite aseptic. The longer, in fact, it is kept, the better it becomes. The gut is preserved for

FIG. 8. Esmarch's and Chiene's method of catgut filopressure.

use in ordinary five-per-cent. oil, which should be occasionally changed, as it sometimes turns rancid.

The use of the catgut ligature and suture is a very important feature of the antiseptic treatment. Experience shows that the gut becomes buried in the tissue, without producing irritation or suppuration, and in the course of some days disappears, being in fact, invaded and supplanted by the living cells of the part. The portion of tissue included in the noose of a ligature does not die, nor does the external coat of the included vessel become divided or ulcerate. The prospects of sound healing of a ligatured artery in a wound are thus materially increased, and there must be less chance of secondary hæmorrhage, while in the ligature in continuity of a large vessel, as for aneurism, the antiseptic noose around the artery produces no ulceration of its external coat and the organisation of the clot within must proceed more quickly and more certainly. There need be less anxiety to avoid applying the ligature in proximity to a collateral branch, and the chances of secondary bleeding are less. The gut may be sometimes too soon converted before obliteration of the included vessel is completed, or the knot when the gut is imperfectly prepared may slip. It is a little less easy to manage, and less firm and secure than silk.

Mr. Chiene has suggested an ingenious way of applying catgut ligatures, which is especially applicable to the smaller vessels, often so difficult to secure. A curved needle, threaded with gut, is passed beneath the bleeding point, so as to include with the vessel a small portion of adjacent tissue; when the ligature is thus tied, it cannot slip, and in aseptic wounds, no necrosis of the included portion will take place. The needle may be armed with a thread long enough to secure several bleeding vessels in succession. It is an application of the method of filopressure. Esmarch in his 'Surgeon's Handbook' figures a similar method of employing the gut.

The gut also answers very well for the purposes of sutures, at all events in wounds where there is no great amount of tension. Gut sutures may be both deep and superficial, and the loops within the tissues should not require to be removed. After four or five days, the dried-up knots may be picked up off the wound-surface, while the portions included in the tissues disappear, and leave no trace behind.

Fine carbolised silk, of the variety called Chinese twist, is often used in preference to gut, being more easily applied and securely tied, especially on small vessels. When prepared so as to be completely aseptic, and used in antiseptic operations, it will not as a rule give rise to symptoms of foreign body. After an ovariotomy, a score or more of fine silk ligatures will be disposed of without difficulty in the peritoneal cavity, and it is generally employed in a thick, strong form for tying the pedicle.

To render the silk thoroughly aseptic and safe to use, it should be boiled for one hour in a five-per-cent. carbolic solution, and afterwards preserved for use in a fluid of similar strength.

Frisch[1] has found experimentally that three hours' boiling is required to absolutely destroy all chance of organisms becoming developed in the silk, but he admits a shorter time would, as a rule probably suffice. A useful form of suture may be prepared by steeping the silk for half an hour in a hot mixture of bees-wax and carbolic acid of a ten-per-cent. strength; the superfluous wax is expressed, and the silk preserved in air-tight bottles.

Silver wire will alway remain an excellent form of suture; as also horse-hair, which is quite non-absorbent, and whose smooth epithelial surface and tough elastic quality make it answer excellently.

Silkworm gut also serves a useful purpose in many cases, but I have found some specimens of it rather brittle.

The careful suturing of the wound is important, in order, so far as practicable, to bring its surfaces into close contact, and prevent any pouches or irregularities into which the

[1] Langenbeck's *Archiv*, vol. xxiv.

wound-secretions might gravitate. In extensive wound-surfaces union is best accomplished by deep, in addition to superficial, sutures. The former are usually wire, but catgut may also be used when there is no great strain. It is better to put in too many than too few stitches, for if the wound be properly drained, the more exactly the parts are brought together, the more speedy will be the healing. No more precise instructions than these can be given, as every case

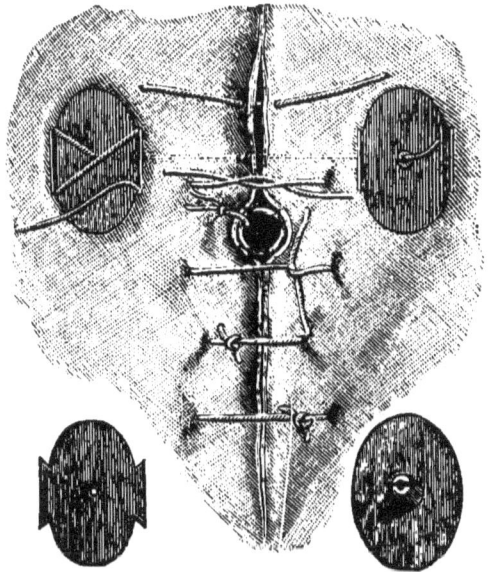

Fig. 9. The leaden-plate suture.

will differ in its details. In passing the deep stitches the needle should be made to transfix the deepest part of the wound, and where deep sutures are used, the drainage must be very carefully attended to. They should only be employed when the approximation of the wound-surface is difficult or impossible otherwise, and they should be retained as short a time as possible when other than catgut, as they tend, by reason of the tension, to excite suppuration.

It is a good thing also to remove the superficial points

of suture early. On the third day, in many cases, the alternate stitches may be cut out, and the rest the following day.

WIRE AND BUTTON SUTURE.

Where a considerable gap exists, as after the removal of a tumour, in which the skin is implicated, deep stitches, for the purpose of drawing the wound-surfaces and margin together with the least amount of tension, are of great

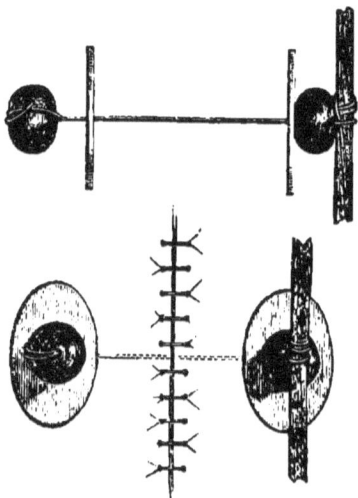

FIG. 10. Professor Thiersch's Bead Suture.

value in facilitating healing, and should generally be used. Silver wire is the material to employ; gut would scarcely answer the purpose.

Various forms of the old button suture may be used with advantage. The drawing displays a modification of it, employed by Professor Lister. A leaden plate, about one-twentieth of an inch thick, is cut in the shape shown in the figure; it has a central perforation for the wire to pass through, and two wings, which can be turned up, and to which the wire may be secured by winding it round in a figure of 8 form. See fig. 9.

The figure represents a wound in diagram, with deep and superficial sutures inserted, and the place shown where drainage tube emerges.

Professor Thiersch has suggested another simple and ingenious method. Outside a leaden plate is placed a small perforated glass bead, which distributes the pressure, and outside the bead the wire is fastened round a small piece of wood, a wooden match answers perfectly, on which the wire is wound, till a sufficient degree of tension is secured. If the wire become loose, one or two further turns will

FIG. 11. Wills' Suture Buttons.

sufficiently tighten it, or the suture can be relaxed, if it be thought desirable. With a piece of sheet-lead, a few beads, a box of matches, and some silver wire, anyone may get ready such sutures with very little trouble.

Dr. Wills[1] suggests silver buttons similar in shape and size to those made of lead: they have two little studs projecting on each side of the central orifice, as seen in the drawings, and round them is twisted the wire in a figure of 8 fashion. The wire can thus with great facility be drawn well home, fastened with ease, and relaxed at will. They are made of different sizes; the most generally convenient size measuring three quarters of an inch in its longer diameter by half an inch broad.

[1] *British Medical Journal*, June 21. 1879.

Drainage Tubes.

Drainage tubes may be made of india-rubber, decalcified ox or horse bone, or ivory; or of metal, as silver, or pewter. Drainage may also be accomplished by a bundle of horsehair, or by strands of catgut; or a slip of silk protective, doubled, and inserted between the edges of a wound, will sometimes answer the purpose, where there is little discharge, and the drainage is required but for a short time.

The efficient and easy drainage of wound-cavities is an essential and important feature of the antiseptic treatment

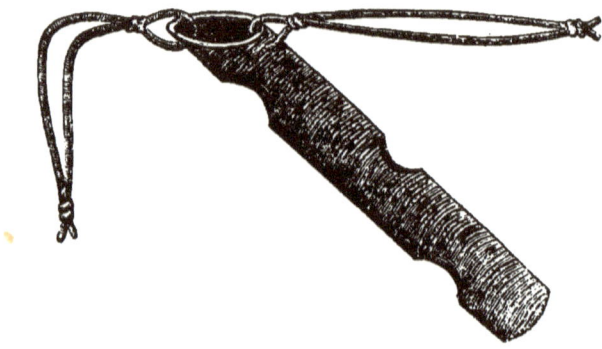

Fig. 12. Drainage tube prepared for insertion.

of wounds. Drainage is intended to prevent the accumulation in any part of the wound of those secretions which almost necessarily occur, and which, if suffered to accumulate, produce tension, pain, and fever, and interfere with the healing process, while in themselves they are most prone to that decomposition, the occurrence of which we know renders the patient liable to very dangerous consequences. Drainage is of the greatest importance during the first twenty-four or forty-eight hours, and in deep, extensive, and irregular wounds.

The more powerful the antiseptic solution employed, and

the more prolonged its employment, the greater will be the amount of after-secretion, and the greater the necessity for efficient drainage. The first dressing is often soaked with bloody serum in twenty-four hours, or even much earlier. Chassaignac, who first methodically employed drainage tubes, passed a long tube from one end of a wound to the other, but his objects were different. Antiseptic drains are short, and when necessary, many in number. When blocked by a coagulum, which arrests their function, they should be taken out, cleaned, washed, and re-introduced, but this is by no means always necessary. In many such cases the tube may be quietly drawn out and the coagulum left behind undisturbed will presently organise. When it is necessary to employ drainage tubes for a considerable time they require

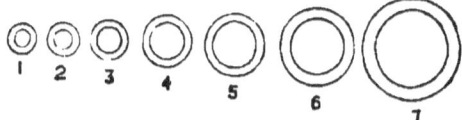

Fig. 13. India-rubber drainage tubing, in seven sizes.

periodical shortening, or must be changed for smaller ones. On the renewal of the dressing they should never be used for the purpose of syringing the wound, so long as it is aseptic. The tube should be large in size rather than small, placed where it cannot be compressed, and have no elbows. The best time to insert the drains is after the sutures have been introduced, but are not yet drawn tight. In large wounds several smaller drains are more efficient than one large one. Two tubes, side by side, often work very well. The time for their removal depends on the amount of secretion. After four to six days the channel in which the drain lies becomes lined with plastic matter, and will remain open for a short time after its removal; where several drains are present they ought to be taken out one after the other, after an interval.

The tubes should vary in size from that of the little finger to that of a quill, and in length from one to four inches; they are perforated with oval openings whose width corresponds to one-third at least of the circumference. The ends may be cut transversely or obliquely, so that they may always terminate flush with the surface, and never project beyond it; any projecting part is pressed on by the dressings, and the other extremity will thus be forced upon and irritate the wound-surface, and the function of the tube is impeded.

Loops of carbolised silk should be inserted at one end, for the purpose of fastening the tube to the skin. They would otherwise occasionally slip into the wound-cavity, and might become healed over, or they might escape externally. Sometimes the external end of the tube is attached by a point of suture to the margin of the wound.

Fig. 14. Lister's Forceps, for introducing drainage tubing.

The fine forceps (fig. 14), is a very convenient instrument for the introduction of the drainage tube and for adjusting it precisely in the place it should occupy.

After being well washed, they are preserved for use in a wide-mouthed bottle, containing a five-per-cent. solution of carbolic acid; or if not kept ready prepared, an assortment of different-sized india-rubber tubes may be preserved in a similar solution in lengths, and cut as required. The rubber seems to a certain extent to absorb the carbolic, and is quite aseptic.

DECALCIFIED BONE DRAINS.

Those at present in use are prepared from sound ox or horse bone, or, still better, ivory. The tubes being first turned of the proper size and length, and placed in a mixture of one part of hydrochloric acid and two parts of water. After ten hours' immersion the earthy matter will have almost completely dissolved away, and the tubes being first washed of the superfluous acid, in five-per-cent. carbolic solution, are preserved for use in five-per-cent. carbolic oil. The width of their lumen varies from three to six mm., and the thickness of the wall from half to $1\frac{1}{2}$ mm. They are two to three inches long.

These drains are the invention of Dr. Neuber, of Kiel, who has published two very interesting papers on the results which may be attained by their employment.[1] The first dressing was allowed in most instances to remain unchanged for fourteen days. After this time, in a large proportion of cases, the wound was found completely healed, except at the points where the drainage tubes emerged, which were occupied in some cases by a superficial granulating surface, which soon healed under some simple application. The strict antiseptic dressing did not require to be renewed—boracic lint, or some similar application, being substituted for it. Thus healing took place under one antiseptic dressing, a most important improvement in antiseptic practice. The drains are absorbed, just as catgut is absorbed or replaced. In six or seven days the drain has become soft and pulpy, and is filled with lymph. After ten days all traces of it have disappeared, except some small portion which may have projected, outside, which will be found lying detached like a small ring and quite unchanged. Thin-walled drains disappear very rapidly.

[1] Langenbeck's *Archiv*, vol. xxiv. p. 314, and vol. xxv. Heft 1.

A safety pin is used to fasten the drain in its place, as silk cuts the material. This, however, is not the case with those made from ivory.

For the first five days, during the period drainage is most required, these drains perfectly serve that object. Then they begin to disappear, and meanwhile do not irritate the surrounding tissues.

The granulations around the drain pierce its openings, fill the interior by degrees, and take the place of its substance, and at last, in a case which has run a normal aseptic course, no traces of it are left.

In wounds which are not strictly aseptic, or where the discharge is profuse, the drain softens down and melts away. Should it happen that the drain lies in the midst of a coagulum, or in gangrenous tissue, it will be found not to be absorbed even at the end of fourteen days. But this non-absorption was observed only once or twice in some 200 cases in Kiel.

The dressing employed by Neuber is the following. No protective or loose moist gauze are made use of, but a pad of carbolised jute inclosed in gauze is secured against the wound by means of a moistened gauze bandage. Over this a second larger pad is superimposed, a layer of oiled paper being contained between the outer surface of the jute and the inclosing gauze. The part is then evenly bandaged all over with an india-rubber roller. When a large pad is to be kept in position such bandages are necessary; they vary from $1\frac{1}{2}$–2 inches in width. This dressing is recommended on the score of efficiency and cheapness, only one dressing is usually required, and the pads are prepared for use by the hospital nurses.

Horse-hair drains consist of bundles of hair, tied together; they may be prepared as follows: According to the thickness needed, a sufficient number of hairs are taken, doubled, and a single hair wound round the bundle in a spiral form to fasten them together, the ends are cut off to the proper length, and the convex loop-shaped end is introduced into the wound; they act by capillary attraction. They should be previously purified of all greasy particles in an alkaline solution, and preserved for use in five-per-cent. carbolic solution. Their employment is only occasional. When in use the size may be gradually diminished by successive removal of some of the hairs.

A bundle of threads of catgut, six, twelve, or more, may be used for drainage. At first they act by capillary attraction, but after a few days, when the necessity for drainage is passed, they become absorbed, and like the decalcified bone drains do not require removal. Catgut may be used to plug the bleeding nutrient artery in a divided bone.

SPONGES.

It is most important to have these clean and aseptic. To prepare new sponges, it is necessary, while they are in a dry state, to beat the sand out as completely as possible, then to wash the sponges in tepid water six or eight times, and leave them to soak, until the water comes off clear and free from sand. Allow them to dry again, and shake them out afresh, to get rid of any remaining sand, and finally preserve for use in a vessel containing five-per-cent. carbolic solution, in which they should be kept at least for one week before using. Getting rid of all the sand is both troublesome and tedious.

The sponges in use during an operation should be washed clean in two-and-a-half-per-cent. carbolic solution.

To purify them thoroughly after use, they should be first washed in water, and then, repeatedly in a weak solution of common soda to remove any blood or matter, then with water again, and preserved as before, in a five-per-cent. carbolic solution. If the vessel be not air-tight, the solution must be renewed from time to time as the carbolic acid volatilises.

If the case have been at all suspicious, the sponges may be steeped for five minutes after the preliminary washing, in the liquor sodæ chlorinatæ B. P. diluted with an equal part of water, which makes the sponge very white and clean.

For ovarian or other operations involving the peritoneal cavity, the finest Turkey sponge should be procured. Eight or ten sponges, about a pound weight, suffice for an operation. Mr. Thornton recommends a good soaking in sulphurous-acid water, which he usually does himself, 1 part of the acid to 15 of water; it whitens and softens the sponge, like the chlorinated soda. They are afterwards placed in 1 in 20 carbolic for some time.

Sponges should never be washed in hot water, in which they shrivel up and become useless.

Precise directions are very desirable with respect to the sponges used during an operation involving the abdominal cavity. Their absolute purity is all-essential. It is necessary to provide against one being left behind, an untoward accident which has happened oftener than once. The sponges to be used, for instance in an ovarian operation, should be counted and laid in a basin of 1 in 20 until the operation begins. Then they are to be wrung as dry as possible out of a warm 1 in 40 solution, and kept by a specially appointed person in a piece of mackintosh which has been previously well carbolised, who hands them to the surgeon as he may require. The 'sister' usually keeps a reserve in case of some unusual demand. The soiled sponges on being handed back to the nurses are thoroughly washed in carbolic solution 1 in 40, then passed through a 1 in 20 solution, and finally wrung once more out of the warm 1 in 40. The sponges used for the later stages

HAND AND FOOT SPRAY PRODUCERS.

of the operation, which have to be introduced within the abdominal cavity, are warmer than those at first provided, and should be as nearly blood heat as possible, but should not exceed it.

When the sutures are about to be introduced the sponges are counted once more to see that the tale is correct. On no account should anyone be allowed to tear or cut a sponge asunder during an ovarian operation. Any missing sponge should be accounted for before the abdominal cavity is closed.

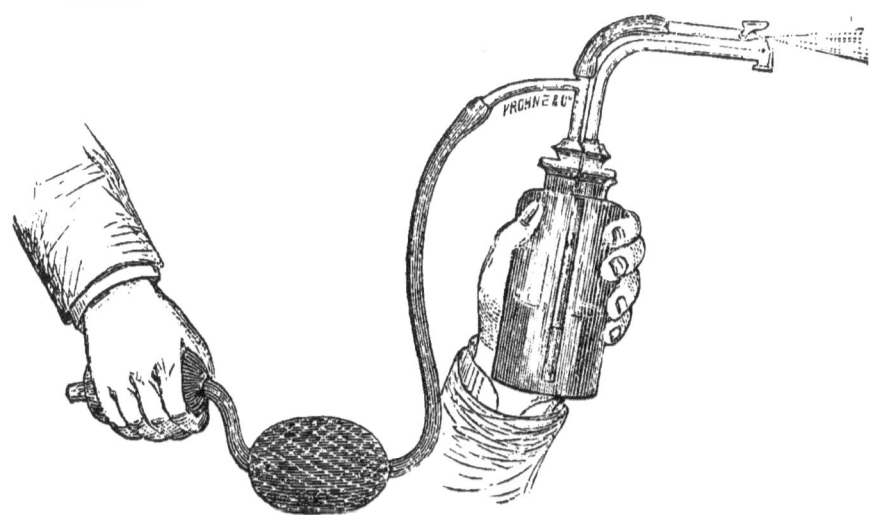

FIG. 15. Hand Bellows Spray Producer.

SPRAY PRODUCERS.

The antiseptic cloud of pulverised carbolic fluid in which antiseptic operations are performed, may be produced by a blast of air, or by steam at a high pressure.

The hand-spray as shown in the drawing, fig. 15, is the earliest form of spray-producer, and the least powerful or perfect. It may, however, be employed in small dressings or for small wounds. The strength of the solution to be used in it is 1 in 40. The spray so produced is rather coarse,

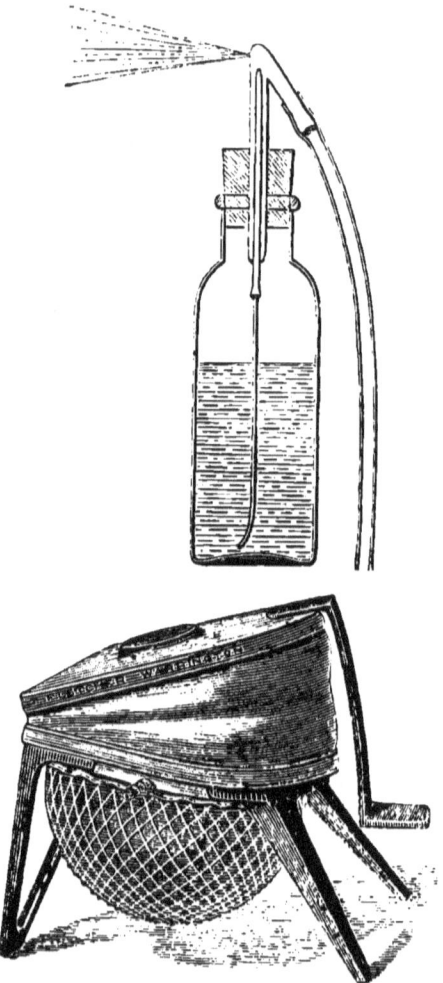

Fig. 16. Richardson's Bellows Spray.

and wets the surface a good deal, which the steam spray with its much higher pressure does not do.

Mr. B. W. Richardson, of Dublin, has devised a more powerful form of spray, worked by a foot bellows, such as is used for the blowpipe in chemical laboratories. That shown in fig. 16 is made by Fletcher of Warrington. It is capable of sending a volume of perfectly pulverised spray a distance of 6 to 10 feet. It may prove useful for dressing, and where a steam spray is not available. The nozzle is of somewhat different construction to the ordinary form. It has a double-action pressure, which is exercised on the fluid in the bottle in addition to the suction.

The steam spray is made of different forms; some kinds with a single nozzle, others with a double, in case one should get out of order, or where it is desirable to cover a wide surface by means of both playing at the same time.

The important points to attend to, are—first, that the lamp shall be of sufficient power, the boiler of sufficient strength, the safety-valve in good working order, and the nozzle of good workmanship and free from obstruction. One-in-twenty solution is placed in the bottle, and the steam rushing over the fine orifice of the tube, leading from the bottle, sucks the fluid up, and disperses it in the form of a fine cloud, the strength of the mingled steam and fluid being at least 1 in 40.

A T-shaped metal mouthpiece filled with sponge is attached to the extremity of the supply tube. It filters the fluid which passes in, and prevents the nozzle becoming choked. The sponge should be removed from time to time, for the purpose of cleaning. Before using a spray producer for an operation or dressing, it should always be examined, to see that the nozzle is free, and that the spray sent out from it contains carbolic. If it be merely steam, the jet will have a bluish colour, the noise as it issues, is different, and the vapour will be tasteless. Temporary compression of the tube shows the difference at once. The carbolic cloud is larger, whiter, and one may taste the carbolic acid in it if the vapour touch the tongue. It should wet the surface very little, and cause no inconvenience when respired. During the dressing of the wound, or an operation, care should be taken to avoid draughts. Some one should have charge of the in-

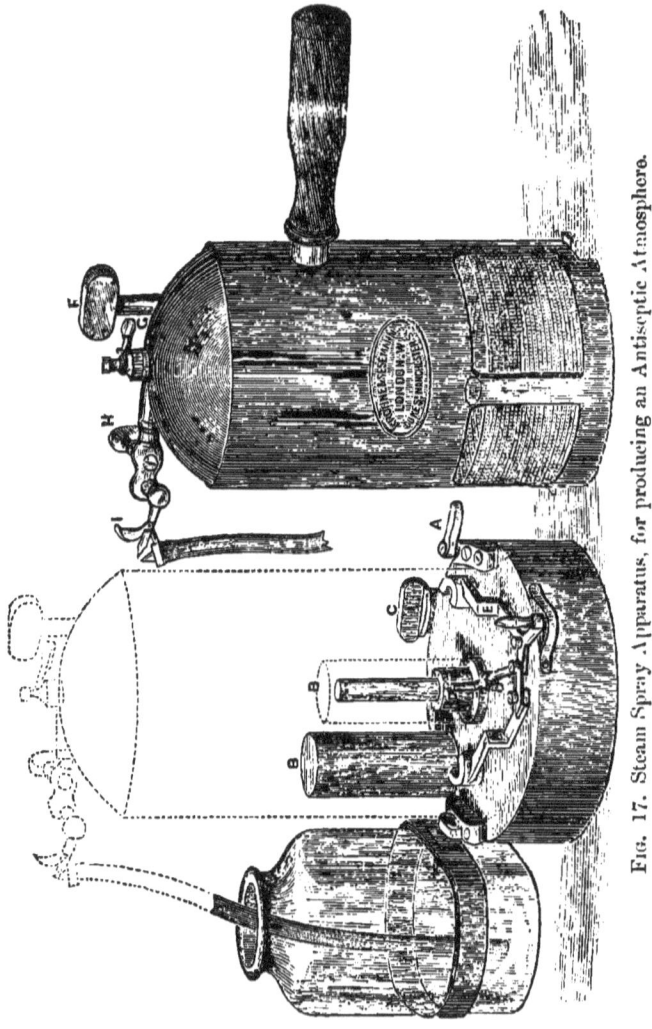

Fig. 17. Steam Spray Apparatus, for producing an Antiseptic Atmosphere.

strument, so that the spray does not, for a moment, cease to envelope the wound. The direction should be such, that the vapour shall get readily under the dressing at the moment it is being removed. If, for any reason, the spray should intermit, an assistant should be prepared at once to cover the wound with moist carbolic gauze. Absolute phenol should generally be used for spray-producing purposes, and invariably in abdominal operations.

The distance of the field of operation from the spray producer is determined by the size and power of the instrument, which should be so placed that the wound will be where the jet is widest and the vapour finest.

The nozzle points of the steam spray are inclined at an acute angle to each other, not as in the ordinary pulveriser, at right angles.

The engraving fig. 17 represents the ordinary form of apparatus. The boiler is removed in order to show the arrangement for lighting and extinguishing the spirit flame. The dotted lines show the boiler in position when ready for use.

The boiler is made of solid brass, to which the various parts are brased, which renders it durable.

When about to use the spray, charge the boiler with clean hot water, allowing for each hour's use about twelve ounces.

Remove the stopper c, nearly fill the lamp with methylated spirit. Light the small wick d, open the clamp e, the small flame lights up the circular wick, the flame of which heats the central tube, and converts the spirit into vapour, which escapes and burns at several small holes near the top of the tube. The cap b, when the apparatus is not in use, prevents evaporation of the spirit. When the lamp is lit, and the boiler charged, a small tongue on the lower part secures the latter in its place.

When an operation is completed the clamp serves to extinguish the central light, the small wick remaining alight. The central wick can be again lighted by opening the clamp, and the pressure of steam needed for another operation is soon got up.

The weight of the apparatus will sufficiently indicate the quantity of water in the boiler. When steam ceases to issue, the lamp should be extinguished; and after use, the boiler should be always emptied, otherwise, in refilling, the exact quautity of water in it would not be known.

Care must be taken to adjust the central wick; it must project evenly, and just far enough to give a small flame around the central tube, because, if the flame be too large, the clamps cannot extinguish it.

Fig. 18. Matthews' Spray Producer.

The wicks have to be frequently changed, as the methylated spirit clogs them when they have been used a few times, and they lose their absorbing power. It may be necessary to pass a fine silver wire through the hole of the nozzles, in order to clear them; the stop-cock of the steam tube should be occasionally taken out, thoroughly cleansed and greased.

The engraving fig. 18 shows a modification of Lister's

spray made by Matthews Brothers of Carey Street. The instrument is carefully constructed; the boiler is of copper, a metal which when tested to bursting point tears and does not fly into fragments. It is globular in form, the two halves being united by hard solder, so that the boiler if made red hot even will not have its integrity impaired.

A beam and weight safety-valve indicates the pressure; its amount is therefore always apparent to those using the spray, an evident improvement on the spring valve, in the previously described apparatus which is liable to get out of order. The spirit lamp is very simple and efficient: an inner tube perforated with holes near the top contains a large solid wick. The size of the flame is regulated by an outer tube sliding up and down the other. When the lamp is fully alight, the flame covers the whole of the under surface of the boiler and an abundance of steam at a high pressure is produced, forming a fine cloud of spray, which easily reaches ten or twelve feet distance. The boiler bears a test label, indicating the pressure it has been subjected to.

Directions for Use.—Boiler to be three-quarters filled with water. Lamp to be three-quarters filled with methylated spirit. When lighted, it should be allowed to burn quietly for a few seconds until the holes round the circular wick have ignited. If the rack become stiff, a little oil should be used to it and to the outside of the tube which works up and down. Neither oil nor grease must be used to stopcock or steam-joints.

This instrument keeps both spray points at work at the same time at a pressure of 60lbs. to the square inch, or one point at 90lbs. pressure. The spray hits at a distance of 20 feet, and lasts from two to four hours, according to whether both points or only one is used and to the pressure employed.

It is difficult to understand how so weak a vapour of

carbolic acid can destroy the power of germs. Possibly it acts mechanically as well as chemically, the septic atmosphere being displaced by the volume of aseptic vapour and fluid issuing from the nozzle; while any germs present, being enveloped in carbolic fluid and falling on a surface impregnated with it, are placed under the most unfavourable circumstances possible for their development. However, this may be, the general employment of the spray appears to have added much to the safety and security of the antiseptic procedure. Prevention is better than cure; hence the protection of carbolised vapour is better than any carbolic irrigation after. The experience of such operations as ovariotomy is sufficient testimony to this. Professor Hegar did not, at the time of the recent publication in Volkmann's *Vorträge*, use spray[1] in his ovariotomies; but if not, he employed elaborate means to purify the air of the chamber in which he operated, by means of chlorine and sulphurous acid vapours. A very small number of persons was allowed to assist, and elaborate precautions were taken to secure their asepticity. Yet, withal, although the cases, 15 in number, were all successful, 6 suffered severely from intra-abdominal abscess, and the progress of others was far from being uncomplicated. His method is, in reality, more difficult than the Listerian, while the results are much behind those obtained by it; for with the spray method, we expect to have little or no fever, and no suppuration whatever, either superficial or deep.

Professor Trendelenberg has recently published[2] a series of cases operated upon with all antiseptic precautions, except the spray, which he omits, with a view of trying to simplify the method, and to render it more applicable to military surgery. He contends that the free

---
[1] Volkmann's Sammlung Hegar, *Uber Ovariotomie*, 1877.
[2] Langenbeck's *Archiv*. Band XXIV.

ACTION OF THE SPRAY.

application of carbolic solution, and thorough disinfection of the wound immediately before the application of the dressings, does away with the necessity for the spray; and he supports his opinion by a number of very successful cases. But this hardly warrants us in abandoning what has hitherto proved itself so efficient a safeguard against septic influences, and which renders the elaborate washing which Professor Trendelenburg advises, unnecessary. It is surely undesirable, if it can be avoided, to employ an excessive irrigation, as it irritates the wound, increases the amount of secretion, and may even sometimes produce carbolic poisoning.

The failure to perfectly disinfect some wounds exposed for a period to the action of the air serves to explain why the subsequent washing out of wounds operated on without spray does not always succeed.

. The spray purifies the air just as rain clears the atmosphere. As Gussenbauer says, it is better to combat the enemy by prevention than to await his invasion, and then drive him out again by the heavy ordnance of repeated irrigation. Those who fail to use the spray abandon the breastplate of their armour, and the protection which it affords against the poison darts in the atmosphere.

ON ANTISEPTIC PRACTICE.

A Sketch of its Objects, with some Hints for its
Application to Various Cases.

The antiseptic method is not intended to supplant or supersede what experience has hitherto proved to be useful in surgical practice, nor does it leave out of sight the fact, that wounds are subject to other than septic influences, severe injury for example, to nerves or blood-vessels, the condition called hæmophilia, the tendency, whatever it be by which tetanus is produced. What we know however, is, that wounds treated in the ordinary way often have a tendency to inflame and suppurate, and may, and sometimes do, end disastrously by blood-poisoning.

The antiseptic method claims absolutely to exclude this hitherto not infrequent cause of death. A collateral but great advantage of the general adoption of antiseptic practice in a hospital is, that even in those cases in which its application is impracticable the risk of wound-diseases is minimised. In fact, the longer a strict antiseptic practice has been carried out, the more perfect does it become, and the excellence of the results increases in proportion as noxious influences become less numerous and less potent.

The antiseptic method of treatment requires a certain amount of previous preparation before an operation is undertaken. Its objects are to procure the speedy and safe union of wounded or injured surfaces, and to avert those risks which are more or less common in other plans of dealing with wounds. This is accomplished by securing a free and perfect escape of wound-secretions, by preventing the possibility of their accumulation and decomposition within the wound-cavity, thereby causing the separation of the wound surfaces, and consequent interruption of healing. Means having been taken to render the wound aseptic in the first instance, putrefactive changes are prevented afterwards, by external antiseptic applications, so that the serious consequences which admittedly depend for their origin on septic changes in a wound must of necessity be averted.

In order to secure these results, everything that may be required during an operation should be ready for use beforehand. The surgeon and his assistants must thoroughly cleanse their hands, and then render them aseptic by washing in a five-per-cent. carbolic solution. The instruments, sponges, and surface of the body on which the operation takes place, must all be disinfected in a similar manner. All parts likely to be included within the area of the subsequent dressing should be disinfected. A layer of moist antiseptic gauze laid on the surface during the time of producing narcosis, is a good plan, or a limb may be enveloped in moist gauze, fastened with a bandage for two or more hours previously, especially if there be sinuses.

For purposes of general description we may suppose an ordinary amputation is being performed, but much of what is said is applicable to any aseptic operation. During the entire progress of the operation, the parts concerned should be enveloped in an atmosphere of carbolic spray, which is most satisfactorily produced by a steam spray-producer.

Narcosis having been induced, and an Esmarch bandage when required put on, the operation may be proceeded with.

During the sawing through of the bone, if a $2\frac{1}{2}$-per-cent. solution be poured over the saw, it will avoid its over-heat ing, which does, of course, tend to some extent to induce necrosis of the cut surface.

For general purposes the following items are the more important, and they should be all prepared beforehand :

One and two gallon jars filled with five-per-cent. carbolic solution.

Two or more small and large basins. Mackintosh sheeting, to protect the patient and clothing from wetting by the spray. It should be washed before using.

The operation-table ought to be quite clean.

Soap and water, spirit, brush, and razor are also to be in readiness.

Flat porcelain dishes, such as photographers use, filled with 1-in-40 solution, are convenient for placing instruments in. These are lifted out as required, and replaced when not in use. Sometimes two dishes are useful, one for those instruments fre-

quently in use, and one for those in only occasional use.

A vessel containing carbolised sponges.

Wound syringes or irrigators.

A bottle of drains, of various sizes and lengths, prepared ready for use in 1 in 20 carbolic.

Instruments as may be required; with clamp forceps in sufficient number.

Strong and weak salicylic wool.

Protective and folds of dressings ready arranged.

Loose gauze.

Ordinary gauze bandages.

Elastic bandages.

Safety pins.

For the success of the antiseptic treatment and the permanence of an antiseptic dressing the complete arrest of hæmorrhage is essential.

When the operation is completed, therefore, every bleeding-point must be carefully secured with catgut, and the ends cut short so soon as the noose is tied fast.

It is convenient to tie the larger vessels first, and then, when all that are apparent are secured, to remove the Esmarch constriction, and without waiting to tie, apply one after another a compressing forceps to each bleeding-point, so that a dozen such forceps may be hanging on at the same time. The ligatures can then be tied subsequently, one after the other.

In an amputation of the thigh, as many as eighteen vessels may have to be secured by ligature; in a leg twelve or more arteries may require ligature. The position of the

CLAMP FORCEPS.

larger vessels is known, and the smaller ones may be generally found in the intermuscular septa.

Mr. Spencer Wells, in an interesting paper, chiefly

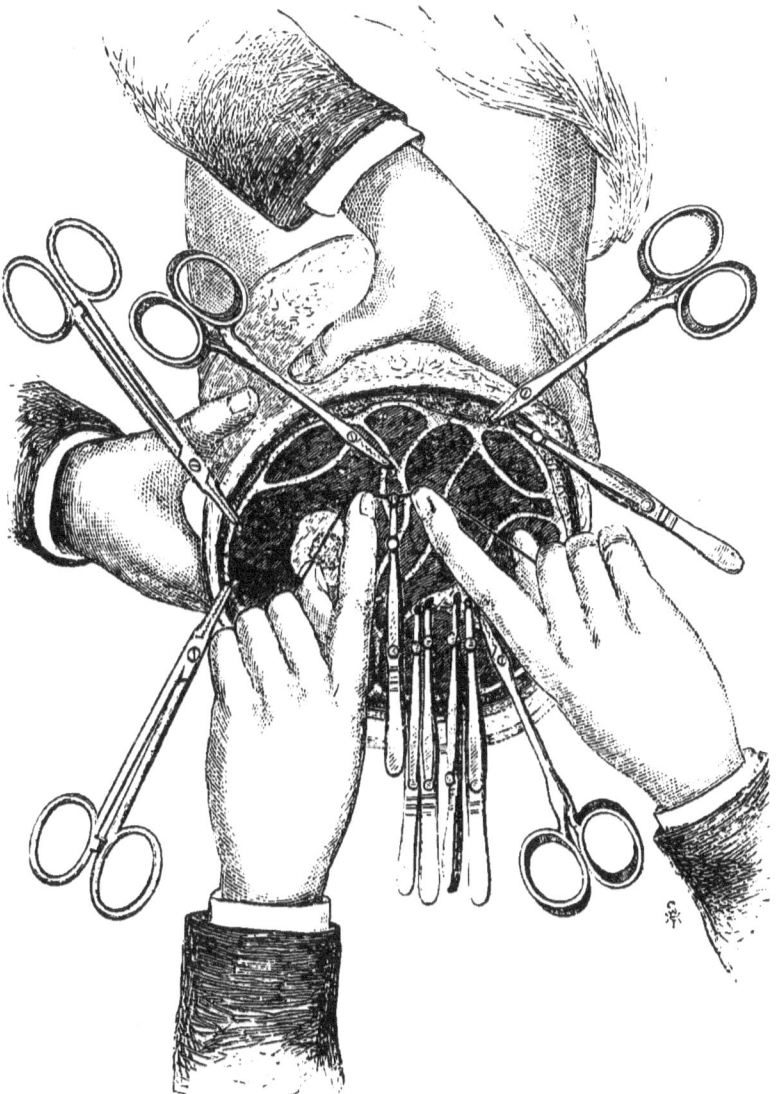

FIG. 19.—Shows the mode of application of the forceps to the vessels in an amputation of the thigh in the middle third.

historical,[1] points out the value of forci-pressure, not merely in procuring a temporary stoppage of the flow of blood, but its definitive arrest in the smaller vessels and veins. He narrates the manner in which he was led from the use of the old 'bull-dogs' to the convenient and powerful clamp forceps he has now employed for the last six years, two of which are represented applied on the left side in fig. 19.

Esmarch, Koeberle, and Pean abroad, and others it may be, have insisted on this valuable method of hæmostasis; but their forceps are inferior in efficiency, and the claims put forth for priority on their behalf in the use of this method of stopping bleeding are scarcely sustainable.

Ovariotomy, as well as many other forms of operation, is greatly facilitated by the use of these forceps, which allow the operation to proceed with scarcely any interruption, whereas much valuable time would be otherwise wasted in securing bleeding-points.

Fig. 20 shows a light, strong, and excellent form of clamp forceps, quite similar in principle to that of Mr. Wells.

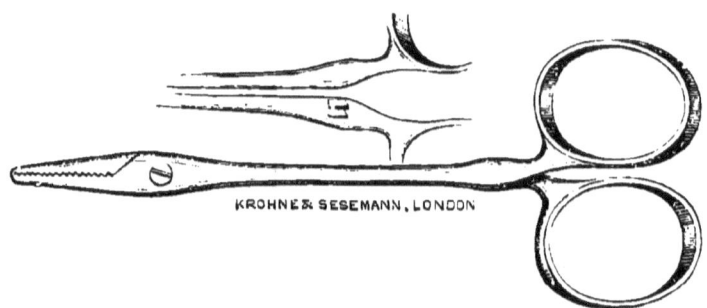

Fig. 20.

Both forms of forceps are seen in use in plate 19, as also the ordinary torsion forceps which may be used for the same purpose.

[1] *British Medical Journal*, June 21, 1879.

The parenchymatous hæmorrhage, frequent after a prolonged use of Esmarch's constriction, may be arrested by cold irrigation with carbolic, or salicylic, solution, and elevation of the stump. Pure spirit is a good application; it is slightly styptic and quite antiseptic.

Hot water is also an excellent hæmostatic, and will sometimes succeed when cold fails; it should be used as warm as the hand can bear, which is about 127° Fahr. It is best applied, not in the form of irrigation, but by means of a mass of folded gauze the size of the bleeding surface, which should be pressed upon it after being dipped into the hot water.

Dr. Hunter has published papers on the subject in the 'Philadelphia Medical Times,' November, 1879. But for many years hot water has been occasionally employed as a hæmostatic in St. Thomas's Hospital. Hunter recommends a temperature of 150° Fahrenheit, and states he has used it as high as 160° Fahr. without producing any ill consequences. This is the temperature at which albumen coagulates; but as the water is intended to act as a stimulus to an increased vasomotor-contraction, and not by a mechanical coagulation of the albumenoids in the blood, this temperature is needlessly high. The application should not be too long continued. A temperature of 110° to 120° will generally check ordinary capillary bleeding.

The bleeding stopped, the sutures are introduced, and so placed as to bring the wound-surfaces as much as possible into contact. Hence in large wounds they are generally alternately superficial and deep (Fig. 21). Catgut may be used where there is no tension, and for superficial stitches. Wire, with lead buttons or leaden plates, may be employed for deep stitches, or when there is any strain to overcome.

The wound is now closed, except at one or more points, where the drainage tubes emerge. These pass from the deepest parts, directly to the surface, and should be cut off level with the skin.

They are placed in the angles of the wound, at the most dependent points, and in any recess or irregularity, where secretions might be confined.

They should not be allowed to remain in contact with the cut surface of a bone.

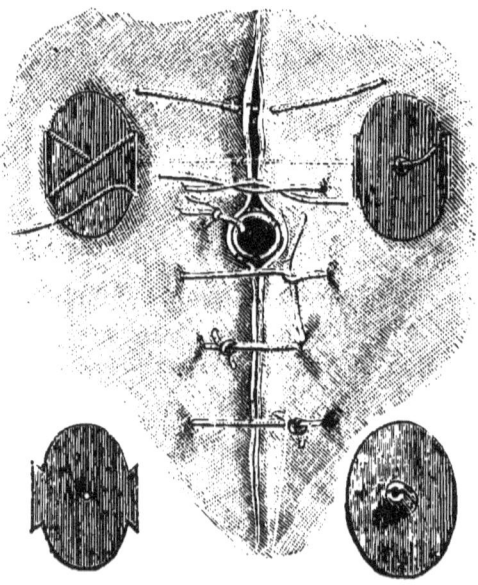

Fig. 21.—Diagram to illustrate the mode of employing deep, superficial, and button sutures.

They can be conveniently fixed by a point of suture or a pin to the margin of the skin, and can be passed with ease into a wound or abscess-cavity with the fine forceps represented fig. 14.

All bleeding arrested, the wound should again be washed with 2½-per-cent. solution to remove all blood-clots before finally closing, so that there need be no subsequent syringing through the drains.

DRESSINGS.

The wound-surfaces may be afterwards gently pressed together by sponges, to expel any remaining blood before applying the final dressing.

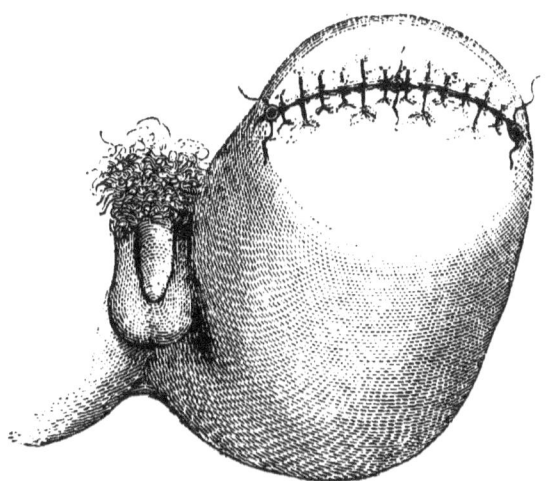

FIG. 22.—Diagram of the stump after an amputation of the thigh. The sutures and drainage tubes are inserted, and the wound is ready for the application of the dressings.

This may now be proceeded with, its function being to continue the action of the carbolic acid in the form of vapour, instead of spray, and to receive and immediately disinfect any discharge from the wound.

First a strip of protective, one half to one inch wide, and extending well beyond the ends of the wound, is applied. This is dipped in the carbolic solution, 1 in 40, immediately before application.

If one long strip do not adapt itself well, two shorter ones may answer better.

Over these a considerable quantity of loosely folded gauze, wrung tightly out of a 1-in-40 solution, is applied in such a way as to press the wound-surfaces together. In the case of an amputation, this pressure should be lateral, and not vertically against the cut surface of the bone. If much secretion be anticipated, it is better to moisten the gauze with 1-in-20 solution. As many layers should be applied as will readily absorb the anticipated amount of discharge, and it is well to bind these firmly to the part with a moist carbolised bandage; thus a firm, equable compression will be exercised as well as thorough protective occlusion. These deep layers of moist gauze are a most essential part of the dressing. Over this it is well to put a layer of salicylic wool, either 4 or 10 per cent., according to the probable amount of discharge, then some turns of bandage to fix it.

The inner bandages are important; they exercise a uniform compression; they keep the folds of moist gauze and salicylic wool *in situ*, and prevent any air getting beneath the margin of the dressing.

The eight-fold layer of dry gauze is now applied with the mackintosh interposed between the outermost layers.

The mackintosh should be dipped, shortly before use, in carbolic solution, 1 in 40.

The gauze must not be spared—either the dry external layers, or more especially the folds of moist gauze first applied.

The external layer should extend on all sides far beyond the wound area; but on the dependent side, the gauze should be most abundant inside, and most extensive outside.

At the margin of the dressing a packing of salicylic wool renders air-exclusion more perfect.

Finally, the carbolised gauze bandages over all procure firm fixation, and should exert a considerable pressure; they are very pliable and mould well to the parts.

When all is finished, the dressing affords support and rest to the wound, and the pressure tends to arrest any oozing of blood.

In arm and thigh amputations, performed high up, in disarticulation at the shoulder and hip joints, the dressing must be secured by a turn of elastic bandage applied at its margins, and in these cases the trunk is included in the dressing. Where the breast or groin are concerned in the wound, the dressing and bandages include part of the corresponding limb (figs. 24 and 48). Whenever there is difficulty in adjusting the dressing, or the part is subject to much movement, the elastic bandage will generally render all secure. Some surgeons prefer to use flannel bandages, in place of elastic webbing for this purpose, as the proper amount of tension of the elastic bandage is difficult to estimate, and the elasticity of the flannel answers sufficiently well.

In many cases, however ample the first dressing may be, another will be required within perhaps twenty-four hours, as the gauze will be more or less saturated with the sero-sanguineous discharge from the wound. The amount of the discharge depends on the extent of the wound, and the degree of carbolic irritation to which it has been subjected, as also on the amount of time and trouble expended in the arrest of hæmorrhage.

The necessity for a change of dressing is indicated by any of the following circumstances:—

1. Penetration of the wound-discharge to the margins or external layers of the dressing.
2. Loosening of the dressing.
3. Severe continuous pain.
4. A prolonged rise in temperature above 100° Fahr.
5. Septic odour.
6. The opinion that the wound is healed or already converted into a mere granulating surface, needing no further antiseptic dressing.

The conditions adverted to in 1 and 2 may be sometimes adequately met by the external addition of some extra layers of carbolic gauze, jute, or bandage.

A second dressing may safely remain two or three days,

and in many cases a week, or even longer; but much longer than one week is not, generally, safe in the ordinary form of carbolic gauze dressing.

After an interval of three or four days, is a favourable period to remove the drains.

The propriety of their removal is indicated by the absence of discharge; the drains are empty and dry, and their removal is not followed by a drop of secretion. Under these circumstances they can be of no further use, their continued presence might excite suppuration, and prevent any speedy closure of the opening left after their extraction. Their interior, too, is sometimes filled with a blood-clot, or a mass of lymph, which remains behind on the withdrawal of the tube.

When all goes well and the dressing remains aseptic, the margins of the wound show little or no swelling. The pain after the first day is trifling. The skin presents a normal colour, there is no redness of the edges, which are firmly glued together; the secretion from the drainage tubes is very small in amount, thin, serous, or merely stained with blood, and it has no odour. There is in short no true suppuration.

Under these circumstances, the less the wound is disturbed and the fewer the dressings, the better.

It is undesirable and unnecessary to syringe out the wound through the drainage tubes. This produces mechanical irritation, it excites suppuration, it prevents union, it entails a longer continuance of drainage, newly formed adhesions are torn asunder, and healing is delayed.

The external surface should be gently cleansed by a stream of carbolic solution from an irrigator or a Higginson's syringe, or by a bundle of wool or gauze steeped in the solution. If any suture is cutting or producing tension, it must be removed at once; but this is rarely necessary, as there is little or no swelling.

At each dressing the progress of healing is of course more or less disturbed, and the spray, which ought always to be employed to maintain an aseptic atmosphere around the uncovered wound, is in itself irritating.

The dressing should occupy, therefore, as short a time as possible, and everything required for it should be ready prepared beforehand.

Fig. 23.—India-rubber syringe for cleansing a wound.

The things in general use are :—

A spray producer.
Porcelain or tin trays, to dip the instruments in.
Basins to contain the carbolic solutions 2½ per cent. and 5 per cent.
An irrigator, or a Higginson's syringe.
Balls of salicylic wool, enveloped in gauze, the size of small sponges.
Scissors and forceps to remove sutures and drains.
Fresh drains, and forceps to introduce them with (fig. 14), in cases where the tube is blocked and any secretion retained, otherwise the tubes should not be disturbed. It is not necessary to remove a tube because it is filled with blood, unless there be discharge accumulating behind it.
Folds of loose moist gauze.
Salicylic wool.
Bandages.
The eight-fold layer, and mackintosh cut to the proper size.

In some hospitals the patient is placed on a small table,

in order to facilitate easy and quick dressing. In a case of a thigh amputation, the sound foot rests on a stool, and the stump projects over the end of the table. By throwing the arms overhead the equilibrium will be preserved. This position leaves the pelvis much more accessible for the application of the bandages and dressings. The parts can be readily cleansed, and the new dressing applied with least loss of time and disturbance. Schede does all his dressings on the lower extremities on such a table. In the upper extremity, the patient may sit on a stool. In this way all wetting of the bed is avoided, which, at the same time, may be remade. A pelvic-support is of use in cases which cannot be removed from bed. It is made of wood, in three different sizes, and the upper surface hair-stuffed and covered with oil-cloth, which can be renewed as often as necessary (fig. 51).

In most cases as soon as the action of the anæsthetic is past, the patient begins to recover, the pain of the wound rapidly diminishes and disappears; a slight temporary rise of temperature, rarely exceeding 99° or 100°, or no rise at all, takes place. There is an almost entire absence of fever, the appetite is normal, and the food, after the first few days, may be the ordinary diet. After an amputation of the upper extremity, the patient would often like to get up the next day. The duration of treatment is curtailed, and the difficulties of nursing are diminished.

The course of the wound after a disarticulation does not differ from what occurs in an ordinary amputation; the troublesome transformation of synovial membrane and cartilage into connective tissue, by granulation, and the not infrequent necrosis of the cartilage, do not take place. We do not know by observation what changes occur in cartilage, under these circumstances; probably, it is gradually replaced by fibrous tissue, and the synovial membrane soon loses its special character.

ABSCESS.

The progress of a case thus treated, compared with the average progress of similar cases, subjected to other forms of treatment, presents therefore a marked advantage at every point.

After an amputation, too, the final result is better; the bone is buried more deeply, and is not so liable to adhere to the cicatrix; the stump is fuller—there is less cicatricial tissue in it; and less tendency to become conical by retraction, or to ulcerate on the surface.

This in general terms is a description of what may be expected, and what we hope to obtain by the true antiseptic management of a large wound, such as that produced by amputation of a limb. The same principles of treatment, and to a large extent the same details, more or less modified, must be carried out in all cases where we wish to reap the advantages of the antiseptic method.

ABSCESS.

In opening an abscess, either acute or chronic, the surface must first be purified, and the incision made under the spray. The point selected should be conveniently placed, both for the purpose of safe dressing, as well as ready drainage. Two or more openings may be made in furtherance of the latter object. The abscess-cavity may be treated in two ways; either without interference at all, or by washing it out. I have frequently washed out the remainder of the pus with a 1-in-40 solution, avoiding over-distension, however. This plan is not always free from risk, and is very likely to be followed by symptoms of carbolic-poisoning, if the cavity be large, in children especially; and it is not necessary when the protection of the spray has been complete, besides directly irritating the lining membrane or abscess-wall. In some cases irrigation may act beneficially, but in most cases it is just the opposite. It is a matter of indifference whether a small abscess be washed out or not.

The drains should be large, in proportion to the size of the cavity, and capable of giving issue to secretions from any part of it. A double-barrelled drain is an excellent form for many cases. In psoas, mammary, or axillary abscess the use of the elastic bandage is required to keep the margins of the dressings closely applied to the surface.

The careful employment of the spray is of great importance at the subsequent dressings of large abscesses, especially those connected with bone. The margin of the dressing should be slowly lifted up, and the spray made to play well beneath it. When the drainage-tube is removed, for the purpose of cleansing or shortening, a fold of moist gauze may be placed over the opening; it affords an additional protection. The frequency of dressing required, depends, of course, on the amount of discharge, but after the first day this is often surprisingly small. Where a joint is concerned, or the bone is diseased, prolonged rest must be enforced, even after the opening is closed.

A large psoas abscess is best evacuated by a small incision. It is useless to make the opening larger than that which the drainage-tube requires for its introduction.

A small incision leaves a small scar—larger wounds become filled with blood or lymph, and afford no advantage in respect of the freer drainage, which it was their intention to provide for before the employment of drainage-tubes became general.

But although under an antiseptic treatment the discharge may be maintained aseptic, a prolonged continuance of it is frequent. The great length of the fistulous tract from the diseased bone in the spine to the external opening renders an effective drainage always more or less difficult.

To avoid these serious objections to an opening made in the groin or thigh in an abscess which has descended thus far, Professor König told me he had on three or four occasions performed the following operation with the most satisfactory results, having previously satisfied himself by experiment that though difficult to do and tedious, the proceeding does not necessarily involve any risk to the peritoneum, or other danger.

The abscess having been first opened below Poupart's ligament, a strong probe is introduced, and on its end an incision is made into the abscess above Poupart's ligament in such a position of course as not to involve the peritoneal sac. From this last wound he was able to introduce a strong, long, probe-shaped instrument, with a handle at one end so as to make it project in the lumbar region.

On this the superimposed tissues were divided on the outer side of the erector spinæ, and the abscess-cavity reached at a point comparatively near or quite close to the seat of the disease. In this manner the immense advantages of direct and complete drainage are afforded. Any sequestra or loose portions of bone can be extracted. In one of the cases a large quantity was thus removed.

The abscess-cavity speedily contracts, the lower openings soon close, and the prospects of a more rapid and complete cure are much enhanced.

In large abscesses, the layers of folded moist gauze should be soaked in five-per-cent. solution before application, and the quantity used should be large in proportion as the discharge is copious.

A large abscess which I recently opened, was in the lumbar region, in a boy. After the first day the discharge was almost nil, and in four or five days the drainage-tube was observed to be filled with grey pultaceous lymph. The tube was removed. There was no further discharge, the fever and pain disappeared after the first day. There was a certain degree of carbolism, but not of any alarming extent. The lad soon recovered.

An interesting case of deeply seated abscess in the pelvis was admitted under my care, February 11, 1880. A boy of eleven, previously in good health, without any known cause, was seized during the night, five days before, with violent pain in the region of the hip and groin.

On admission, he was suffering intense pain, and in high fever, the temperature 102·6°.

On examination, a deep-seated, hard, extremely tender and tense swelling was felt in the left iliac fossa. There was no fluctuation, and no change in the superficial parts. Both the general and the local symptoms were so severe that I felt called upon to attempt to give relief. I made an incision parallel to and a little above Poupart's ligament, about two inches in length, and divided seriatim the layers of the abdominal parietes, until I exposed the transversalis fascia, just as in the operation for ligature of the iliac artery. A deep search had now to be made, the parts held up towards the median line, until at last, with the aid of the director and finger, an abscess-cavity, containing three ounces of pus, was discovered in the deepest part of the wound, and evacuated. It lay within the true pelvis, behind the body of the pubes and between the side of the bladder and the obturator internus muscle. The drainage-tube introduced to the bottom of the abscess from the surface was quite five inches long. Immediate relief was given, the swelling fell speedily, and the boy made an uninterrupted recovery. The most careful examination to discover the cause of the suppuration gave a completely negative result.

The case is quoted to illustrate the advantage of early evacuation of pus in such cases. Bound down as it was

COMPOUND FRACTURE.

by reflections of pelvic fascia, and so deeply placed, fluctuation could not be felt, and it is difficult to say how far the suppuration might have spread, or what damage it would have occasioned, had not timely assistance been rendered.

The antiseptic dressing of the wound in this instance required great attention. The use of the elastic bandage to

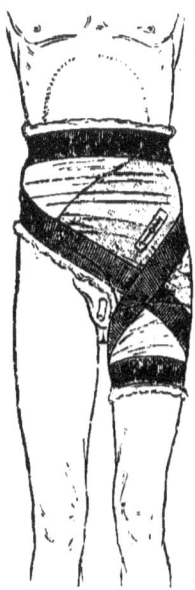

FIG. 24. Antiseptic dressing as applied in this case, with the position of the wound indicated.

secure its margins was essential. In fig. 24 a sketch of the dimensions and appearance of such a dressing is given, taken from the subject of the previous remarks.

In strumous abscesses, incise—evacuate the pus, and use the sharp spoon, to remove diseased granulation-tissue, or an infiltrated lymphatic gland, then drain. The granulation-tissue in these abscesses will not heal, and contains, in all probability, tubercle. So long as it is left it acts like a foreign body, and keeps up irritation and suppuration. It should be removed just as carious bone is removed.

Fistulæ should be laid open freely, and the sharp spoon used to remove the unhealthy granulation-tissue lining them. This practice may be extended with advantage to fistulæ connected with joints, and even to the diseased joint-cavity itself in young children. Good results may be obtained in this way, without proceeding to the more formal and formidable operation of excision.

Compound Fractures.

In no kind of surgical injury have the results accomplished by the antiseptic method been more thoroughly satisfactory and complete than in compound fracture. In future we may expect to save the limb of the patient in all cases in which the extent of damage to the soft parts, vessels and nerves, is not such as to absolutely forbid the attempt. Even in cases where the expectation of saving the limb is not great, we are justified in giving the patient the benefit of the doubt, as we do not endanger his life by so doing ; and should gangrene or any necessity for operation occur, we may then amputate without increased risk.

Pyæmia and septicæmia, which have hitherto caused half the deaths among the fatal cases, should no longer occur; but only death from unavoidable causes, such as fat-embolism, delirium tremens, tetanus, senile bronchitis—causes not directly dependent upon the wound, nor its treatment.

The first thing in dealing with a case of compound fracture is thoroughly to cleanse the entire limb with soap and water, certainly every part of it which may be included in the subsequent dressings, and afterwards wash it with strong carbolic solution. An anæsthetic is necessary, as the manipulations are often very painful and protracted. The wound must in recent cases be thoroughly irrigated with two-and-half-per-cent. solution, or if it have been exposed for several hours, with a five-per-cent. solution. If the external wound be too small to allow of the most complete irrigation, it should be enlarged. The disinfection of the whole wound-cavity is the most important part of the practice, and one or several counter-openings may require to be made in order to wash and completely disinfect every recess. It must be thoroughly well done. In some of the slighter cases the external wound may be closed; but more frequently drainage-tubes must be inserted, and the fractured limb having been adjusted, and antiseptic dressings applied, it is at once put up in a suitable splint or apparatus.

In those cases where the wound-cavity is large, or when it has been exposed for some hours before the patient has come under treatment, the washing out must be repeatedly made with the stronger carbolic solution, and drainage will be invariably necessary, as the amount of secretion afterwards is considerable. Drainage is also required in all cases where the tendency to bleed is not completely checked. In short, it must be employed in a very large majority of the cases. When the bone has pierced the skin it should be disinfected in the same way as the soft parts. Necrosis does not appear to follow any the more on this account. Only when a fragment interferes with the 'setting' of the fracture, or is completely detached, ought it to be removed. Any extensive removal of bone is prejudicial and unnecessary, for even very loosely connected pieces will re-unite; while an extensive removal of bone renders delayed or non-union very likely to occur. Counter-openings are made with a double object. When the wound-cavity is large and irregular they facilitate its complete disinfection in the first instance, and its thorough drainage afterwards. There should be no hesitation in making them either of sufficient number or size, where the seat of fracture cannot be completely disinfected through the original wound. It is undesirable to pass a drainage tube between the fragments of the fractured bone unless drainage cannot otherwise be satisfactorily accomplished. In the instances, however, in which this has been done it has not induced necrosis, an occurrence the probability of which under any circumstances is greatly diminished by the antiseptic method.

In recent compound fractures, therefore, the external wound will often require to be freely enlarged, and counter openings made, to completely disinfect, and efficiently drain, the seat of injury.

The ends of the bone may be turned out, examined, and, if necessary, any sharp points denuded of periosteum, or irregular edges removed with pliers, or completely detached fragments extracted.

The entire wound-cavity is then washed out with the 1-in-40 or 1-in-20 solution for about 15 minutes or longer, according to the circumstances already detailed, drainage-tubes in sufficient number are introduced, the skin sutured where it has been incised, up to the margin of the tubes, the usual dressings applied, and the splints outside of all. The dressings must be sometimes changed on the following day, on account of hæmorrhage, or excessive secretion, but except on these accounts the wounds in many cases heal primarily, and in most cases very few dressings are required; in the most favourable cases only one. The first dressings need not be disturbed for three or four days; then the drainage-tubes may probably be removed, after which the wound will speedily close.

The subsequent dressings required may be applied at intervals of five or seven days. The dressing in other cases will not need removal for a week or even fourteen days, especially in cases where drainage is either not required, or after the drainage-tubes have been removed, or where the absorbent bone-drains have been used. Then when the fracture is first examined the wound is often found either quite superficial or completely healed, when of course the fracture may be dealt with as if it were an ordinary simple one.

The aseptic course exercises an important influence upon the torn tissues; the muscles are not damaged by inflammatory infiltration and suppuration, so that their subsequent repair is more speedily and perfectly effected. Their healing resembles that of the muscles torn during a dislocation, the dissection of a recent case of which shows how extensive this damage may be without any permanent injury to the subsequent function of the joint.

It is alleged, but I think without sufficient evidence, that the antiseptic method favours the occurrence of delayed or non-union. It is not apparent why this should be so, any more than it is in cases of simple fracture.

A certain number of cases of compound fracture with a small external wound, may be 'sealed' as heretofore. A pad of antiseptic dressing, either gauze or wool, is applied externally, and the wound heals under a scab.

There is always some uncertainty as to the selection of such cases, and the result must be always doubtful, but when the conditions are favourable, and little or no septic matter has found an entrance, recovery will take place, just as it has hitherto done, in all respects as in simple fracture.

Plaster-of-Paris bandages are in many instances the fittest form of apparatus to employ in combination with antiseptic dressings to give the needful support to the limb. They must be prepared so that they may be removed and reapplied in their integrity. Cutting a window in an ordinary plaster bandage does not give sufficient room for the application of the antiseptic material, while it weakens the splint so much as to render it untrustworthy as a support.

An arrangement for treating simple fracture, which I have used for some years, is applicable to cases of compound fracture of the leg, or with appropriate modification to those of the thigh also. Strips of coarse blanket flannel cut to a suitable length and breadth are employed after dipping in plaster of Paris cream, to form anterior and posterior splints.

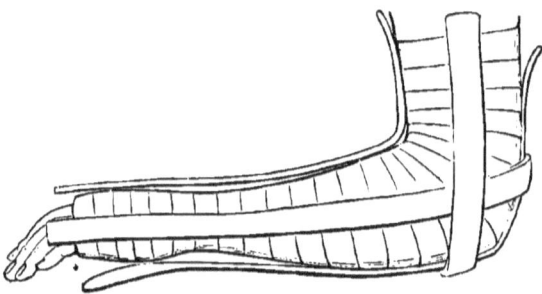

FIG. 25.—Illustrates the manner of strengthening a plaster of Paris splint with strips of thin wood. (Esmarch.)

Care should be taken that they are narrow enough when applied to the limb not to overlap at the margins.

A convenient and manageable splint for fracture below the knee may be made of four such strips, see fig. 26. These should be long enough to extend on the anterior and posterior surface of the limb, from the toes to above the patella; 2 being applied in front and 2 behind. They are secured in place by circular gauze bandages which may be plastered or not according to the strength needed, and the likelihood of an early removal being required. Sometimes three layers of flannel may be applied posteriorly, or the strength increased by interposing strips of pasteboard or of thin wood (fig. 25).

Figs. 27 and 28 give the shape of these flannel splints after they have been moulded to the limb. If carefully prepared and not exposed to wet they will last during the whole treatment of the case. Impregnation with wax or, still better, brushing them over with a saturated alcoholic solution

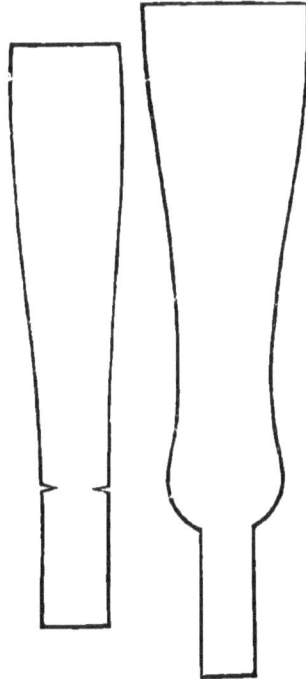

FIG. 26.—Shape of the anterior and posterior flannel strips. They are cut sufficiently long to extend from mid thigh to the toes, and drawn of the shape best suited for an average limb. The rounded expanded lower part of the posterior splint fits the heel.

of soap will greatly increase their strength and durability. The apparatus complete with a telegraph wire introduced

FIG. 27.—Anterior Leg splint.

into the anterior half of the splint for the purpose of suspending the limb, is represented in fig. 29.

The limb must be held in proper position until the plaster sets; a process considerably hastened by applying a supplementary dry bandage outside, which may be afterwards

Fig. 28.—Posterior Leg splint.

removed. In simple fracture the splints are in direct contact with the skin, but in compound, the dressing will be interposed.

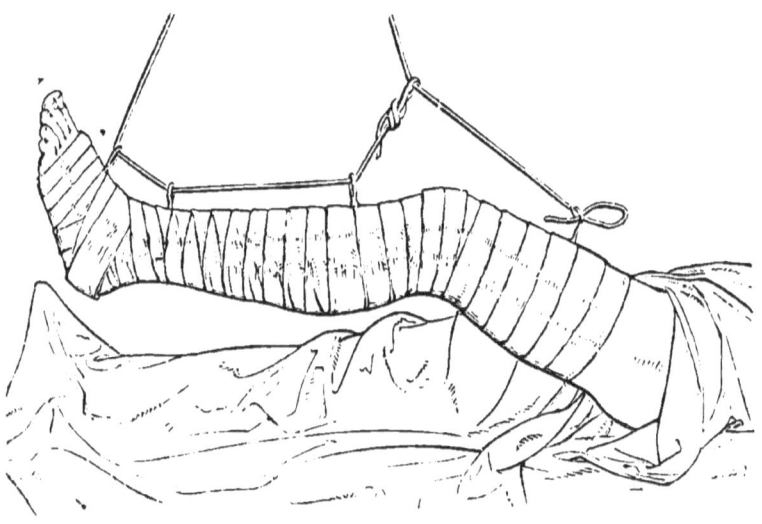

Fig. 29.

When the wound is on the anterior aspect of the limb, as it usually is, and there are no counter-openings behind, it is not difficult to arrange the posterior splint, so that it shall not require removal except when there is much bleeding or discharge. It should be made comparatively narrow, fig. 30, lined with mackintosh, and a layer of folded gauze laid over it; the external dry dressing may be applied outside this splint, and inside the anterior one. The anterior splint can be lifted off, the dressing renewed and the condition of the wound inspected without any disturbance to the fracture, as the limb remains supported by the posterior splint, fig. 31,

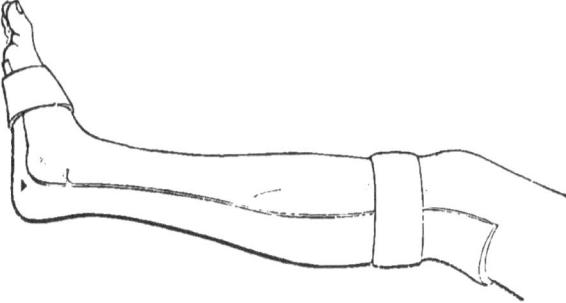

Fig. 30.—Another form of posterior plaster-of-Paris splint.

out of which the limb can, however, be lifted in case blood or discharge have reached it. The mackintosh lining, however, to a large extent, prevents the posterior splint being damaged. Without additional trouble the anterior splint can be applied so as to remain in place while the posterior

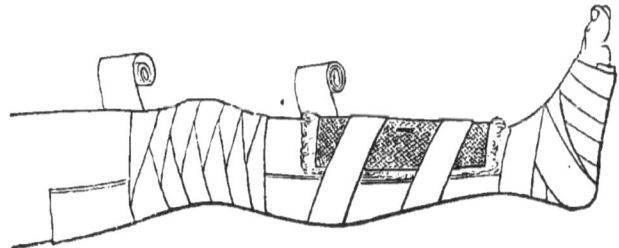

Fig. 31.—Dressing for a compound fracture of leg.

one is removed, when the wound is on the posterior aspect of the limb or there are counter-openings.

I find that Dressers seldom know the best way to prepare plaster of Paris for use. It is often made too thick, and sets too soon; or, it is too thin, and additional plaster is added at the last moment, which makes the mixture lumpy and unmanageable. A sufficient quantity of water for the purpose in hand should be first poured into a basin, and then the plaster lightly shaken into it, handful after handful, or spoonful by spoonful, but *without stirring*, until the plaster just begins to float on the surface of the water; then enough of plaster has been added, and, on stirring, it will quickly blend with the water, and a homogeneous mixture, of the proper consistence, that of thick cream, will be the constant result.

In this the flannel strips are dipped, and they will take

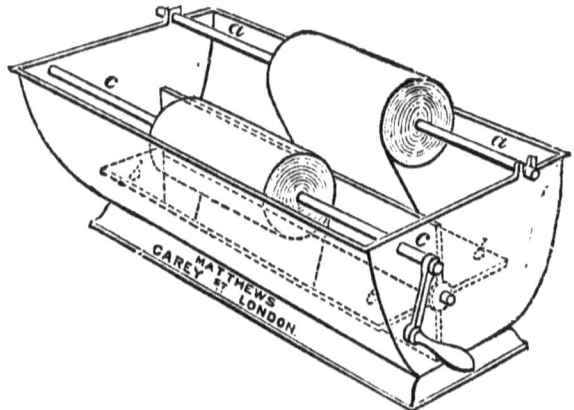

Fig. 32.—Dr. Walker's roller, for preparing bandages with wet plaster of Paris. The trough is filled with plaster of Paris, mixed with water in proper proportion. If some gum be added, it retards the setting and gives more time for the necessary manipulations.

the mixture better, and form a stronger and more durable splint if they have been previously wetted, all superfluous water being wrung out. Thus encased, it is very easy to suspend a limb by loops of bandage to a bed-cradle, or by means of telegraph wire rods introduced into the substance of the splint it may be hung up to a support above the patient's bed (fig. 29).

The bandages used to fix and strengthen the splints may be prepared with dry plaster well rubbed into the meshes, after which they are rolled up and kept ready for use in an air-tight box. Before application care must be taken to soak them sufficiently long in water. It requires a few minutes for the fluid to penetrate the interior of the bandage, or bandages may be passed through a plaster cream in a basin just before they are required, and applied in the moist form, which makes a stronger, more durable kind of bandage. The apparatus for preparing moist bandages shown in fig. 32, answers very well, especially for those who are careful about soiling their fingers.

Plaster-of-Paris splints are so generally useful, and applicable to so many cases, that I have thought it desirable to describe a convenient way of making them.

OSTEOTOMY FOR RICKETY DEFORMITY, PSEUDARTHROSIS, AND IN FAULTY UNION AFTER FRACTURE.

I have in another place alluded to a series of about 50 cases of division of bones with a chisel or saw for the purpose of removing deformity without a single ill result.

Of 182 cases of osteotomy for rickety deformity, collected by J. Boeckel (including 30 previously unpublished cases of Volkmann's), not a single death occurred, and in all the deformity was removed.[1]

A few fatal cases have occurred from septic poisoning, and some joints have become anchylosed after suppuration.

Still the almost complete safety of these operations on bone now appears to be demonstrated, while we can recollect, just as in the case of compound fractures, how very serious and frequent were the bad consequences of such operations on bone previously, especially in the lower limb.

We are now justified in performing many operations for deformity which we previously would not have attempted.

The knock-knee of young persons can be treated with

[1] J. Boeckel, *Nouvelles considerations sur l'ostéotomie dans les incurvations rachitiques des membres.* Paris, 1880.

great facility in this way, and with I think less suffering and inconvenience, and certainly less loss of time, than in the splint treatment, which is often difficult to manage, worrying to the patient and to his attendants, and frequently quite ineffectual. In young children, and milder cases, I often employ repeated forcible straightening under chloroform, putting up the limb afterwards in plaster-of-Paris splints, for the deformity can be often rectified by such means. But in severer cases of pronounced genu valgum,

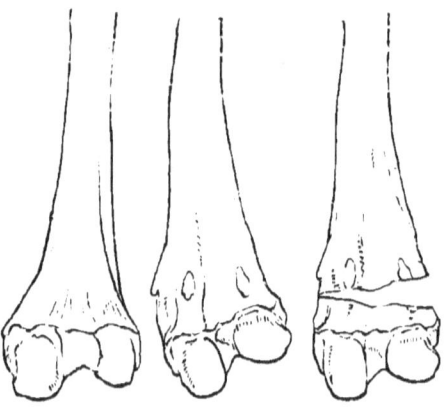

Fig. 33.—Shows the normal femur compared with one in which the lower end of the diaphysis is elongated by increased growth on the inner side. The line of junction of the epiphysis and diaphysis is indicated in the drawing, and was plainly visible in the actual specimen, from which it is apparent that the change in length on the inner side is not in the condyle but in the shaft (Mikulicz). The third cut represents the position and size of the gap in the bone required to effect straightening the limb. This deficiency becomes subsequently filled with bone, so that the defective development of the femur on the outer side of the bone is compensated for, and its effect removed.

or varum, anterior curvatures in femur or tibia, the simple section of the bone with a chisel is easy, rapid, and safe, while splints and apparatus are inefficient. Cuneiform resections are seldom necessary.

The saw is an instrument I rarely use, and in cases of knock-knee I have for the most part abandoned Mr. Ogston's most cleverly devised oblique division of the internal condyle.

I now prefer to divide the femur transversely, entering the chisel on the outer side of the limb through a small wound made for the purpose. The advantages of this method are that the mechanism of the joint is not interfered with, as it is in Ogston's operation, an interference in some instances followed by stiffness, while in most cases a considerable period elapses before the joint-mobility is restored, whereas mobility is immediately restored after the other plan on the splints being removed. The basis on which Ogston's operation is founded, viz. that the deformity in

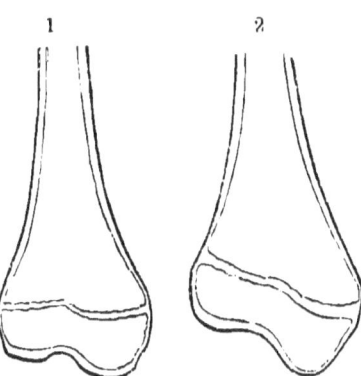

FIG. 34.—1. Section of a normal femur in a well-developed boy of sixteen. 2. Section of femur taken from a young carpenter of eighteen, with right genu valgum (Mikulicz).

knock-knee depends on an unnatural elongation of the internal condyle, is, I think, successfully disputed by Mikulicz[1] for at all events many of the cases, so that the *raison d'être* for Ogston's method ceases to exist. The deformity depends rather on an excessive growth in the lower internal part of the shaft, the thickness of the epiphysis itself remaining unaltered; as the figs. 33 and 34, accurately re-

[1] 'Die seitlichen Verkrummungen am Knie,' Langenbeck's *Archiv*, Band xxiii. Heft 3.

duced from Mikulicz's full-size plates, appear conclusively to prove when the normal and diseased conditions are compared together. At other times the tibia is mainly at fault.

It is not difficult to compare the size of the two epiphyses in a case of single knock-knee, from the fact that the tubercle for the insertion of the tendon of the adductor magnus corresponds to the epiphysial line, and can be readily felt. The distance of this bony prominence from the joint surface of the internal condyle can be measured on each side in the

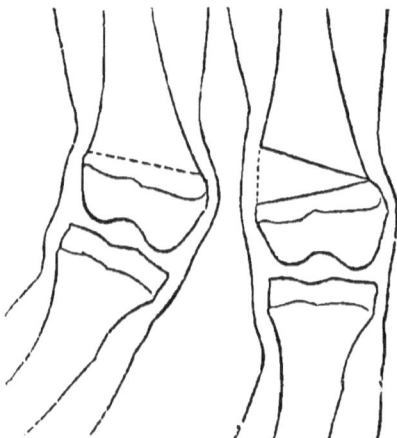

FIG. 35.—Diagram intended to show the relations of the parts before and after the operation for *genu valgum*.

bent position of the knee, and the difference, if any, ascertained.

I have therefore in all recent osteotomies for knock-knee divided the femur with a chisel just above the epiphysis, making the section run nearly parallel to the articular surface. In this way, about three-fourths of the thickness of the shaft may be divided, the remaining fourth being broken during the straightening of the limb. The wedge-shaped gap left subsequently becomes filled by bone-callus, and the limb is thus slightly lengthened.

The operation having been completed with the usual antiseptic precautions, a plentiful dressing is applied, the limb put up in plaster-of-Paris splints, and not disturbed for two or three weeks. At the end of this time, when the splints and dressings are removed, the wound will usually be found healed, and the bone-union firm. Another, or the same splint, is applied for a further period of a fortnight, when the patient will be well.

The small wound made for the chisel is left unsutured, so that any oozing of blood may more readily escape. Drainage is not often required. When the wound is large, or the oozing considerable, a drain may be inserted, and if it be of decalcified bone the dressing need not be disturbed for its removal. In none of the cases operated upon has there been constitutional or local disturbance. The patients recovered without more inconvenience than the necessary restraint caused by the splints occasioned. In double cases I have always operated on both sides in immediate succession upon the same day. The plaster-of-Paris splint is strengthened by the application, on the outer side, of long splints bent so as to form a slight concavity. The two splints are afterwards bracketed together across the chest and over the ankles with iron hoops, which keeps the patient from twisting about in bed.

Fracture of the Patella.

The security afforded by the antiseptic method opens up a large field for operation in cases of ununited fracture. The most novel and daring application of the method has been to cases of ununited fracture of the patella. These fractures, we know, often unite faultily, that is, by a feeble membranous connection. A very short thick fibrous bond of union is as good as a bony one. But as a rule this is not obtained, and the treatment required to obtain it extends over a period of from three to six months.

Dr. Hector Cameron of Glasgow, first, then Mr. Lister, Mr. Rose, Mr. H. Smith, Professor Esmarch, Professor Trendelenburg and others, performed this operation, not only in cases of old fracture, but in recent ones. They have laid bare the fragments, opened the knee-joint by an anterior longitudinal incision, and after removing blood and clots and washing out the joint-cavity, sutured the fractured surfaces together with silver wire. The sutures are inserted so as to

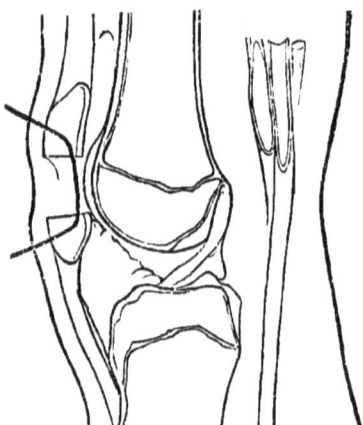

FIG. 36.—Vertical section of the knee joint, illustrating the operation for an ununited fracture of the patella.

be extra-articular, the articular lamella and cartilage being avoided. Fig. 36 illustrates the relations of the parts concerned in the operation and the position of the sutures.

After the bone-surfaces are approximated, the wires are twisted, the ends are cut off a quarter of an inch outside the level of the skin, the wound is then united, and a drain inserted. A drain must be also inserted through an opening at each side into the knee-joint. The wire sutures may remain four weeks or more, or until firm union of the fragments takes place. A considerable number of cases have been now recorded in which good bony or very close fibrous union was obtained without serious symptoms of any kind occurring during the recovery of the patient. Passive movement should be resorted to at an early period, three to four weeks, else the joint is very apt to remain stiff.

Professor Trendelenburg narrates [1] a case of fracture of the patella in a man of seventeen years, of eight weeks' standing with an inch separation. He made a curved incision above the patella, bared the fragments, which were found osteoporotic, and introduced three silver sutures. The fragments were drawn together with some difficulty, one of the sutures breaking out. The periosteum was united by catgut stitches. Only three dressings were required—on the second, eighth, and fourteenth days. The sutures healed in, the result was excellent, the joint-mobility free, and the bone-union all that could be desired.

In old-standing cases much difficulty has been experienced in drawing down the upper fragment, and the quadriceps tendon has had to be divided.

A proposal made by Schede [2] seems an excellent one and the practice preferable for recent cases, in which certainly one of the chief difficulties of obtaining good union arises from the effusion into the joint interfering with the approximation of the fragments. They are kept apart by it during the first eight or fourteen days, the period during which such important processes for the repair and future consolidation of the bone are taking place in cases of fracture. As a means of obviating this hindrance to proper fixation of the fragments, Schede would puncture and wash out the joint with antiseptic precautions; a proceeding of late so decisively shown to be free from danger. Of five cases treated

[1] *Verhandl. Deutsch. Gesellsch. für Chirurgie*, vol. vii.
[2] *Centralblatt für Chirurgie*, No. 42, October 20, 1877. p. 657. Schede on the treatment of transverse fracture of the patella.

in this manner, in three a firm bony consolidation was obtained; in the other two, a fibrous union with close approximation. It was only, he states, in consequence of these two patients having been allowed to make too early a use of the joint that bony union was not in their case also obtained.

The mode of procedure is as follows. The joint is first punctured with a full-sized trocar, also, if needful, the prepatellar bursa. The still fluid mixture of blood and synovia escapes readily through the canula, and through this a three-per-cent. carbolic solution is injected until it escapes clear and all blood is completely washed out. The canula is then removed, and the opening covered with a piece of protective silk, over which a ball of salicylic wool is fixed. The fragments can now be accurately coapted and placed in good position; they can be thus fixed by strapping. The whole limb is then bandaged with a flannel roller, figures of 8 being carried around the knee to help the action of the strapping, and lastly a plaster-of-Paris bandage extending from ankle to hip is applied to the fully extended limb. The strips of strapping must be renewed within eight days, as they otherwise become ineffective, by reason of the further subsidence of the swelling around the joint, and they often need to be again renewed at the end of a further week or ten days. Schede keeps on the plaster-of-Paris splints for six weeks, after which the treatment consists in preserving the newly formed cicatrix in the bone from stretching, particularly bearing in mind the greater strain upon it at first in consequence of the contracted condition of the Quadratus femoris, the origin and insertion of which have been approximated as closely as is possible during the whole duration of the treatment. Gradual and cautious flexion should only be permitted during the first four to six months, the amount being by degrees increased at monthly intervals. The mobility of the joint may be promoted at a later period by passive movements, 'massage,' baths, and douches.

In some of Schede's cases he did not wash out the joint with carbolic acid solution, but a wider experience leads him to recommend that this should be now done in all cases.

In one case also he made use of Malgaigne's hooks with a surrounding dressing of salicylic wool. He considers that, with the precaution of having the points first gilded, they may be used under such a dressing without any fear of ill effects in the way of irritation and inflammation.

BADLY UNITED AND UNUNITED FRACTURE.

The correction of viciously united fracture may be more readily undertaken in cases of union with deformity after fracture of the shafts of the bone in the leg, thigh, forearm, or arm. When the hip or knee is anchylosed in an inconvenient position the limb may be made straight by division of the bone with the chisel near to the former joint, or in severe cases by the resection of a wedge-shaped portion of the bone.

In ununited fracture the risks incident to exposing the fractured ends, resecting and suturing them no longer hamper the operating surgeon.

Mr. Lister has published twenty-six operations of this kind, including eight in the thigh and nine in the leg, without a fatal result in any. Some of the cases were very difficult; all were serious operations. In the thigh solid bony union was obtained in six of the cases, a splendid result. In one, a fracture of neck of femur, the union was only fibrous, but the patient could subsequently walk without the aid of crutches.

An objection urged against the antiseptic method in the treatment of these cases of non-union, is that the local reaction is so slight that it does not sufficiently provoke the reparative processes necessary for the consolidation of the fracture. To obviate this difficulty it has been suggested to remove the antiseptic dressing after some days. Thorough fixation of the fragments is, I think, the most important part of the treatment, and in furtherance of this the ends may be

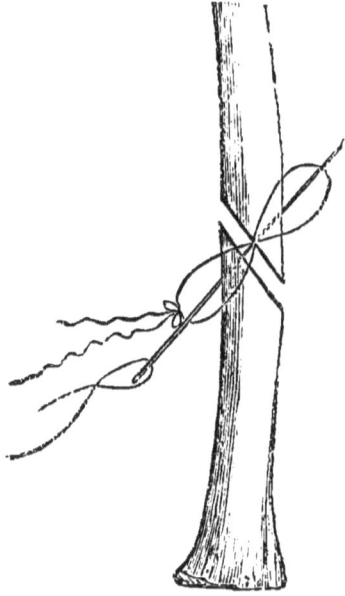 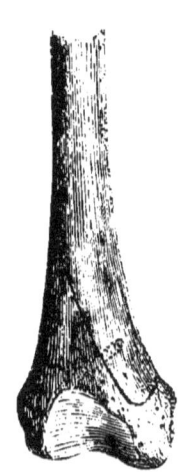

Fig. 37.—Suture for an ununited fracture.

Fig. 38.—The woodcut is copied from a drawing which Professor Trendelenburg sent me, showing the line of fracture in his case and the position of the ivory peg used to fasten the fragments together.

fastened together by ivory or metal pegs being driven through them, after drilling a hole for the purpose, or by sutures passed through holes drilled to receive them, or by fastening the fragments with a silver wire passed around a needle transfixing the bones, as in the simple arrangement shown in Fig. 37 which I have borrowed from Mr. Mason. When a sufficient time has elapsed the needle is withdrawn and the loop of the wire comes away with it.

A much more frequent employment is now made, however, of the old ivory pegs of Dieffenbach, and iron pegs or even screws are often employed for the purpose of more securely fixing the fragments.

I have used pointed iron nails in this way with the best results in delayed and non-union. In the external wound made to gain access to the ends of the broken bone no inflammation was occasioned, while sufficient disturbance was excited to procure union, but this is unfortunately not invariably the case.

Trendelenburg [1] reports an interesting example of a cure in this way of a case of pseudarthrosis of the lower end of the right femur. A man of fifty-six years sustained a very oblique fracture of lower end of the right femur passing through the knee-joint. After nineteen weeks' treatment a pseudarthrosis was found to exist. On flexing the knee, the apex of the upper fragment became very prominent close to the inner border of patella, and moved freely backwards and forwards within the joint-cavity. The knee-joint was laid open by Langenbeck's resection incision, which is a long curved wound on the inner side of the knee. The patella having been displaced outwards, the finger could then be placed in a cleft between the fragments which was partly occupied by the synovial membrane and capsule, the interposition of which had no doubt prevented union. This material was dissected out, the operation occasioning such free bleeding as to need as many as fifty ligatures for its suppression; the two fragments of bone were then drilled and fastened together by an ivory peg driven through both (fig. 38).

Afterwards the knee-joint was thoroughly washed out,

[1] *Verhandl. Deutsch. Gesellsch. für Chir.* vol. vii.

drains introduced, and the wound sutured. The result was most favourable; the wound healed without fever or local reaction; one month later an abscess formed at the back of the thigh, possibly caused by a small sequestrum. But after being opened it gave no further trouble. The bone united firmly, and the subsequent motion in the joint was good. A small protuberance could be felt over the superficial end of the ivory peg.

Langenbeck and Lister have practised somewhat similar operations on the neck of the femur when it remained un-united. The former transfixed the trochanter detached head and portion of neck with a drill, and then fastened them together with a gimlet-like screw. Langenbeck's case was fatal, the joint was penetrated, and it was done before the days of antiseptics. The operation was performed on an old woman with an oblique fracture of the neck partly extra- and partly intra-capsular; in which a painful pseudarthrosis had formed. The operation was followed by inflammation of the joint, and the wound becoming infected with hospital gangrene, the woman died. König has repeated Langenbeck's operation with a silver screw in a young patient and obtained a highly satisfactory result.

Lister's patient recovered fairly well; but, as the two bones had not been pegged together, union did not occur.

The fragment of head is very difficult to keep steady during the boring, and to pierce it at the proper spot from without inwards is far from being easy. Trendelenburg has suggested a daring proceeding to meet these objections. He proposes to make a short incision directly over the neck of the femur, resembling that made by Langenbeck for excision of the hip-joint. The seat of fracture will in this way be at once and easily exposed; the limb is then forcibly adducted so that the trochanter may be bored from within outwards; the limb having been replaced, the drill is introduced through the outer end of the hole, the broken-off head steadied in position with the forefinger, and the drill then worked into it. The fragments are now united with a silver screw, and the wound closed after the introduction of adequate drains. The screw may be removed after fourteen days; a metallic one is necessary, as ivory pegs or screws are not strong enough to sustain the weighty lower extremity in position.

ANTISEPTIC PUNCTURE AND WASHING OUT OF DISTENDED JOINTS.

The large joints may be punctured when distended by blood, by chronic synovial or inflammatory effusion, by catarrhal purulent fluid, or by pus, and their interior washed out with strong carbolic solution without risk to the patient, or *quoad* the operation, any injury to the functions of the joint. A full-sized trocar, such as is used for puncture of the abdomen, should be employed. In some cases, the wound is at once closed; but in others, especially cases of chronic effusion, a re-accumulation of the fluid is to be apprehended, and permanent relief is more likely to be afforded if a drainage-tube be inserted. This allows the fluid poured out by the serous membrane lining the joint, as the result of the irritation caused by the carbolic injection, at once to escape. Where a drainage-tube is required, a small incision must be made to provide for its easy introduction into the joint-cavity.

Rinne points out [1] how completely harmless the antiseptic injection of the knee-joint is. He expresses the conviction that this method of treatment is reliable both as to rapidity of cure of the disease in many kinds of cases and the prevention

[1] F. Rinne: 'Die antiseptische Ponction der Gelenke und das Auswaschen derselben mit Carbolsäurelösungen, *Centralbl. für Chirurgie*, No. 49, Dec. 1877, p. 793.

of recurrence, and that it is likely to come into general use in cases of joint-affection unaccompanied by serious lesion of bone or cartilage.

The desirability and propriety of adopting this means may be considered in regard to the following classes of cases:

In acute synovitis, where an excessive effusion either threatens to rupture the capsule, or causes great suffering; or in those subacute and chronic cases where the effusion becomes stationary, and is not lessened by pressure or other treatment. Puncture should also be resorted to in cases of hydarthrosis, where previous means of cure have proved themselves ineffectual, and in the cases which form the transition stages between hydarthrosis and pyarthrosis—the catarrhal suppuration of joints of Volkmann.

In acute hæmarthroses, the result of trauma, puncture may be usefully practised; it prevents the development of the hydarthrosis, not uncommonly resulting from effusion of blood into joints. As already pointed out, it greatly favours the successful treatment of transverse fracture of the patella.

Certain cases of arthritis urica, with much effusion, may be punctured because of the relief given by drawing off the effusion.

In cases of suppuration of the knee-joint dependent on inflammation of bone it may be employed as a palliative means; or as an adjuvant to other methods of treatment. In the slighter forms of fungous joint-inflammation, puncture, if frequently repeated, may have a beneficial effect.

The operation is performed in the following manner. If it be the knee the exterior of the joint is first well washed, and then under the spray a full-sized trocar is passed in on the outer side. A stream of carbolic acid solution (three- or five-per-cent.) is then injected into the joint, and allowed again to escape, the joint being manipulated meanwhile, so as to ensure the penetration of the solution into every recess. This is repeated until the fluid which flows out is quite clear. Sometimes as many as ten injections are needed. The canula is then carefully withdrawn, and an occlusive dressing applied. A splint is put on externally, and the patient removed to bed. In five or six days the wound will be healed, and the dressing may be removed. If there is no re-accumulation of fluid, a flannel roller is applied, and after some days' stay in bed the patient is allowed to get up with a splint on to prevent movements of the joint. If the joint keeps right, passive movements may be shortly commenced, and in favourable cases the cure is soon completed.

It may, however, be retarded by the re-accumulation of fluid.

I have had good reason to be pleased with the result of puncture and washing out of the knee-joint in some cases in which I have resorted to it. I introduced in one case a decalcified bone-drain with the happiest effect. It allowed much fluid to drain away during the first three days, and then it disappeared, having apparently melted down. The case was one of catarrhal purulent effusion in the knee after a confinement. The patient experienced speedy and complete relief; the fever the woman was suffering from, and the severe pain, at once abated and speedily disappeared, and she made a rapid recovery.

In an old man of 68, admitted with acute suppuration in both the knee and shoulder joints of the right side, coming on suddenly with high fever and a general typhoid condition resembling the pyæmic state, but not associated with any previous injury, I made three large incisions in the knee, allowed the pus, which had already burst the capsule, to escape, and washed out the joint with five-per-cent. carbolic solution. Drainage-tubes were passed across the joint, and obliquely upwards from each side through an opening made at the top of the superior synovial pouch.

The shoulder joint was treated in a similar fashion. The fever at once abated. The man, who appeared at death's door, immediately began to improve, his convalescence being only delayed by an attack of suppurative bronchitis. The wounds in the knee and shoulder soon healed, and he has two fairly movable joints. The case is one of remarkable recovery in a patient of such advanced age, with so severe a malady.

In the suppurating joints of strumous children, much, I think, may be effected by laying open the abscesses around the joint, and the joint itself, thoroughly removing diseased granulating tissue with the sharp spoon, disinfecting them with a five-per-cent. carbolic solution, and after an antiseptic dressing applying an apparatus to secure immobility. I think that in many cases the formal resection of the articulations hitherto practised will be superseded by a trial, in the first instance, at all events, of this less serious procedure.

ABDOMINAL SECTION.

In every text-book it is laid down how very serious penetrating wounds of the abdominal cavity are, and properly so, for death but too often follows. Year after year modern surgery has begun more decisively to approve of laying open the abdomen to remove ovarian and other tumours, or a diseased spleen or kidney, to relieve an intussusception, or an internal strangulation, to evacuate an echinococcus cyst of the liver, or concretions from the gall-bladder, or a calculus in the kidney.

In many cases of intestinal obstruction which would have been treated by injections and opium, or in which an artificial anus would be made in the loin or groin, an opening which rarely if ever heals, we now incise the abdominal cavity, and try permanently to remove the actual cause of the condition.

Gastrotomy and gastrostomy, performed in the one case for foreign body, in the other for disease, are not either uncommon, or, considering the circumstances, unsuccessful operations.

Czerny has made successful experiments in the dog to prove the possibility of excision of portions of the stomach and of the pylorus. About half the cases of cancer of the stomach affect the pyloric end, and a third of these reach the post-mortem table before the disease has contracted adhesions to surrounding parts. So it would seem that where an exact diagnosis is possible our surgical resources may extend to the excision of a cancerous pylorus.

That such operations have been performed under antiseptic precautions with a degree of safety previously unapproached can only be denied by those who choose to ignore the ordinary rules of evidence.

Two great risks during exposure of the peritoneal cavity consist in the rapid cooling of so great a surface, and in its intensely active absorbent power. Its irregular surface affords many recesses for secretions to accumulate in, as at the reflections at the borders of the diaphragm, at the kidney, or the pouch of Douglas; and if the conditions admit of decomposition taking place, rapid blood-poisoning soon follows, and nowhere of so acute a type. Besides, a further and extreme danger arises when there is a wound of any of the abdominal viscera, by which blood or intestinal contents may escape in greater or smaller quantity into the serous cavity. When an operation involves the opening of the cavity of the intestinal tube the greatest care must be taken that no fluid or solid matter escapes from it into the abdominal cavity, or into the layers of the abdominal wall.

OPERATION FOR THE CURE OF FÆCAL FISTULA.

This distressing condition generally results from neglected strangulated hernia and gangrene of the intestine in consequence. Dupuytren's classic operation for the division of the valve-like eperon which forms in these cases, and obstructs the passage of fæces, very often fails to effect a cure, while it exposes the patient to the danger of a severe or fatal peritonitis. At best it is an uncertain and imperfect method. Billroth, Czerny, and others, have devised a more complete form of operation, and one which, with the protection we now possess against surgical accident, cannot be esteemed so dangerous as the former method with the enterotome-clamp of the great French surgeon, while the result is much better for the function of the intestine. In a paper recently published on this subject, Schede [1] relates the history of three cases of artificial anus which were closed by means of an incision into the abdominal cavity combined with excision of a portion of the gut.

The artificial anus in all three cases resulted from hernia; the first was femoral, the second umbilical, and the third a hernia in the linea alba below the navel.

These cases were unfitted for the ordinary operation with Dupuytren's clamp; in one because it was impossible to make out the direction of the lower opening, while absolutely no fæces escaped per anum; in another because of a large ulcerated surface around the mouth of the fistula, and its thin lips; while in the third case no eperon existed, and a spontaneous cure seemed very unlikely. The first and third cases were entirely successful, while the second unfortunately died in consequence of an embolus, derived from

[1] Max Schede. Ueber Enterorrhaphie.—*Verhandl. Deutsch. Gesellsch. für Chirurgie*, 1879, viii. p. 78.

an old thrombus in the femoral vein, becoming impacted in the pulmonary artery.

Schede performed the operation in the following manner.

The patient so far as possible completely abstained from food for twenty-four hours, the gut was completely cleared of fæculent matter by purgatives, and enemata were afterwards administered through the artificial opening. The finger was now passed into the fistula in order to determine the position of the upper portion of intestine, and in a direction corresponding to this an incision was made through

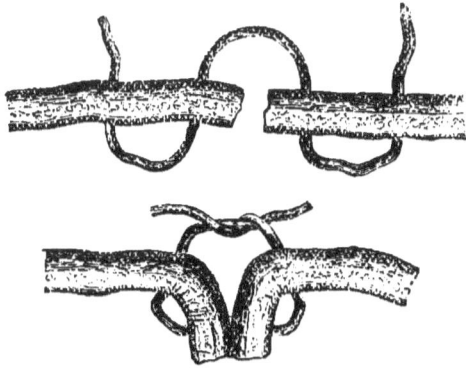

FIG. 39.—Lembert's intestinal suture. The drawing represents a section of the large intestine about twice the natural size. The upper is the peritoneal surface, the lower is that of the interior of the tube.

the abdominal wall, layer by layer, until the gut was exposed, not extending, however, quite into the fistula. A temporary ligature of thick catgut was now passed round the gut and tied with sufficient tightness to prevent any escape of the intestinal contents during the later stages of the operation. The small portion of gut below the ligature was then disinfected by five-per-cent. carbolic lotion, and the remaining bridge of skin between the incision and artificial anus cut through so as to entirely free the end of the intestine. The lower end of the bowel was discovered imbedded in adhesions, it was freed in a similar manner, and a ligature was passed round it at some distance from the opening. The distal and proximate ends of the gut were now 'refreshed,' and the margins accurately united with sutures. Lembert's mode of suture was used (fig. 39), a wedge-shaped piece of mesentery corresponding to the absent portion of intestine excised, and the temporary ligatures which had thoroughly succeeded in their object were then removed, not a single drop of intestinal contents having escaped. The abdominal wound was only partly closed in the first two cases, a large enough opening being left to allow the loop of gut to be secured outside the abdominal cavity by a long needle passed through the edges of the abdominal wound and the mesentery. In the first case the needle was removed on the tenth day, when the gut was spontaneously retracted. Its surface was granulating, and the edges had completely united together. In the second case, which died suddenly on the fourth day, the gut had also healed completely.

In the third case the adhesions were so numerous that the intestine could only be detached for a very short distance. Schede returned the bowel in this instance after stitching the edges together, and completely closed the abdominal wound. In consequence of some fever and vomiting he removed two sutures the next day lest any secretion might be retained. Afterwards the patient made a rapid recovery.

Strict antiseptic precautions were observed throughout in all the cases, and in the first two the loop of gut remaining externally was preserved from pressure by a bird's nest of gauze, a piece of protective having been first applied. Opium was given internally, and the nutriment was fluid and as scanty as possible. Schede draws attention to the great advantages gained by the temporary ligature of the gut, as it prevents any escape of intestinal contents during the operation. He also raises the question of the propriety of at once returning the intestine. His third case was successful, but in his first case, also successful, some fæcal matter was noticed to have escaped into the dressing on the sixth day. He points out, however, that if the gut be kept outside the abdominal cavity till the cut edges unite, it loses the advantages it might otherwise gain by adhesions to neighbouring parts.

Esmarch, in a similar sort of case, in which he replaced the gut at once, lost the patient on the sixth day on account of fæcal extravasation taking place.

RADICAL CURE OF HERNIA.

I have referred to this question in the body of my Address. The operation has again become of prominent interest in consequence of the brilliant results obtained by antiseptic treatment in other procedures involving the peritoneal cavity.

With a view to eliciting the opinions and experience of other surgeons Schede[1] has published 8 cases in which a radical cure was obtained by somewhat different methods of operation in each.

The first case was one of double reducible inguinal hernia, which in consequence of the large size of the abdominal rings could not be commanded by a truss. The operation was performed by exposing the neck of the sac, freeing it from its connections, drawing it down as far as possible, applying a ligature to it, then dividing the neck and removing the sac below the ligature. On one side a piece of omentum was adherent; the pedicle of this was included in the ligature, and the remainder removed with the sac. This man left the hospital in two months, able to follow his employment, with

[1] Max Schede: 'Zur Frage von der Radicaloperation der unterleibsbrüche. *Centralblatt für Chirurgie*, No. 44, November 1877, p. 689.

both rings closed, although they were previously large enough to permit the introduction of three fingers.

Schede next had a series of 3 cases in which the operation followed on herniotomy.

1. Femoral hernia, strangulated for five days, a man, æt. 76. Excision of the sac, and ligature of its neck was practised as in the last operation. The healing was protracted, but the wound closed at end of two months. This patient died of an internal complaint four months later. A slight depression of the peritoneum was then found at the internal orifice, which was closed with a firm cicatrix, so that the finger could not be pushed into the crural canal.

2. A female with hernia above the centre of Poupart's ligament, strangulation for five days. Sac opened, and afterwards excised in the same manner as before. The wound completely healed in twenty-four days. Hernial ring remains firmly closed.

3. A female with irreducible femoral hernia of eight years duration, strangulation for twenty-four hours. Sac opened, small loop of gut returned, adherent omentum removed, sac then dissected out, drawn down, and cut off as before, cure in twenty-four days. Crural opening firmly closed, no impulse to be felt on coughing.

In the next two cases of strangulated hernia this mode of operation was not feasible.

1. A female with strangulated oblique inguinal hernia. Some gangrenous omentum was adherent to the neck of the sac which was full of putrid fluid. The sac was laid open throughout its whole extent, and a drainage-tube inserted into the abdominal cavity. The cavity granulated under antiseptic treatment, and in forty-three days the patient was cured by the obliteration of the sac.

2. A female, æt. 34, with a congenital inguinal hernia; strangulation for two days, caused by a band passing

across the mouth of the open processus vaginalis. Some omentum in the sac was cut off, and its pedicle fastened in the mouth of the canal by a suture. The lower third of the wound was closed. In the upper two-thirds the serous membrane lining the vaginal process was sutured to the skin, as in Volkmann's operation for hydrocele, and the inguinal canal was left open with a drainage-tube in it. Lister's dressing was applied with pressure exerted in such a way as to bring the lips of the wound against the posterior wall of the canal. In thirty days the cure was complete, and the canal obliterated.

The eighth case was one of irreducible inguino-scrotal hernia, complicated by a large hydrocele of the cord, in a male, æt. 53. The rupture had existed for twenty-seven years, and had been irreducible for the last six months. The patient suffered from chronic bronchitis and emphysema. A radical operation for hydrocele was first performed, and four months later a radical operation for hernia. The sac was very thick and intimately adherent to the skin. The intestine had contracted such numerous adhesions that it took some two hours to dissect it out. The intimate connection of the sac with the skin and the stiffness of its walls made the ligature of the neck impossible, and the dissection out of the sac unadvisable. So a similar procedure was adopted to that made use of in the case of congenital hernia, and the patient was completely cured in one month.

In 3 of the cases the operation was performed on account of the patient being unable to earn a living.

In all, the primary object was to procure an obliteration of the sac.

In cases of operation for strangulated hernia the proceeding does not increase the risks of the operation.

The difficulty of applying a perfect antiseptic dressing in this region is very great, both with regard to exclusion of air, and the prevention of contamination by urine. As to the final result, in the majority of cases the operation can only be expected to put the patient in such a position that by wearing a truss the future prolapse of the hernia will be prevented.

The most recently devised operations aim in part at a narrowing of the neck of the sac; and in part at a total obliteration of the sac, while some methods attempt to realise both combined.

Czerny objects to subcutaneous suture on the ground that one does not always know what may be included. Steele first utilised Lister's dressing for the operation; he exposed the ring, and sutured the pillars together. Cure took place in three weeks, but a recurrence of hernia happened six months afterwards.

The operation for removing the sac is an old one successfully revived. It was resorted to on several occasions between 1858 and 1861. Schmucker practised it, and the elder Langenbeck and others have had good results after ligature of neck and excision of sac. Czerny [1] and Risel [2] consider it essential to combine suturing of the pillars of the ring with the ligature and excision of the sac in order to obtain a satisfactory result.

OVARIOTOMY.

In ovariotomy cumulative experience proves that the best results are unattainable without antiseptic protection; that carbolic acid does not irritate the peritoneum, nor the contents of the abdominal cavity; and that carbolic poisoning is rare. The argument that ovariotomy without Listerism

---

[1] 'Studien zur Radicalbehandlung der Hernien,' *Wiener medicinische Wochenschr*, Nos. 21 and 24, 1877.
[2] *Deutsche medicinische Wochenschrift*, Nos. 38 and 39, 1877.

is already so successful that it is needless to employ it, is negatived by the fact that in the hands of most experienced ovariotomists a series of many deaths may occur in succession, and that nearly all the fatal cases are due to some form of septic poisoning.

The practice of ovariotomy has upset many old notions. It has shown that the peritoneal cavity may, under favourable circumstances, be opened with impunity; and that, so far from traumatic peritonitis being a likely result, the great serous surface of the peritoneum is very tolerant of injury. Then it came to be recognised, pre-eminently through the teaching of Dr. Marion Sims, that the frequently-occurring peritonitis after abdominal section was septic in its nature; and that the symptoms produced were those of blood-poisoning. So soon as this was admitted, and the antiseptic method applied to guard against its causes, the operation at once became comparatively quite safe in the hands of all surgeons who believed in the possibility of warding off septic influences during and after the operation.

The few who previously practised ovariotomy with great success, did so by the minutest attention to every detail which might ensure cleanliness and purity, and they attained, as in the case of Mr. Wells and Dr. Keith, a marvellous success. But that success is transcended by their later experience since both these surgeons have adopted complete antiseptic precautions. And in the hands of general surgeons the mortality has been reduced, from a proportion of not less than 50 per cent. all round, to very small figures indeed. Olshausen mentions his results in a paper published in Volkmann's 'Klinische Vorträge,' which may, I think, be fairly taken as typical of those achieved by many other surgeons before and after the introduction of the antiseptic method Before the introduction of antiseptics he had operated on 40 cases, of which 19 died; since, he had had 19 antiseptic cases with but one death. Not only the greatest amount of success as regards safety to life, but the most perfect kind of success in respect of easy and quick recovery, is attained by those who employ the strictest antiseptic precautions in all operations which involve opening the peritoneal cavity.

The operation, in hospital practice, should be done in a small ward, in which the patient subsequently remains. It should be disinfected beforehand with chlorine vapour.

All possible septic influences in the clothes or person of the patient, and in everybody who assists at the operation, must be kept away.

The hands of all concerned must be washed in 5-per-cent. carbolic solution. The instruments are laid in a $2\frac{1}{2}$-per-cent. solution.

Soap and carbolic lotion are used to wash the abdominal surface. The hair is shaved off the pubes.

A double spray should be employed, and the bottle filled with warm 5-per-cent. carbolic solution, which helps to avoid the too great cooling of the abdominal contents.

A frequent washing of the hands during the operation is necessary, as well as each time before introducing them into the belly. A basin for the purpose should be at hand, the fluid in it being changed as often as it is soiled.

The sponges used at the operation should be prepared as has been previously described. Their proper cleansing during the progress of the operation is of great importance.

The intraperitoneal ligature of the pedicle with silk or catgut has proved a great advance in the practice of ovariotomy. The previous antiseptic conduct of the operation renders this method the safest and the best. The pedicle is tied in two or more portions, according to its size, and the abdominal wall closed completely over it. There is no risk of the death of the portion included in the noose, and little of its causing an abscess; nor does it appear to excite any after-trouble when silk ligature is used, in preference to gut, a material hitherto generally rejected, for this special purpose, as being less manageable and less trustworthy.

When the pedicle is secured, the cavity of the peritoneum is cleared of all blood-clots, or fluid which has been extravasated during the operation. And, although this should be carefully and completely done, it has been found unnecessary to do it to so large an extent when operating with the spray, as it certainly must be done when the spray is not used. Olshausen in 17 antiseptic cases removed only the blood and cyst contents lying on the surface. All recovered but one, which proved fatal from embolism.

The sutures are introduced so as to include the entire thickness of the abdominal wall, and the peritoneal margins as well. This brings the edges of the deep part of the wound into contact. The deep sutures thus inserted are generally of carbolised silk; but they may be of wire, or even gut, as there is no tension of the abdominal wall— afterwards as many superficial gut stitches as are needed are interposed between. The employment of drainage is seldom required. What little effusion there may be is aseptic, and becomes rapidly absorbed without causing mischief.

When the operation is complete, and the wound closed, several masses of moist gauze are placed on the belly, over a narrow strip of protective, and a moist bandage is applied to keep them *in situ*. The inguinal regions, and also the pubes and epigastrium, should be well padded with salicylic wool, and another bandage applied over this; then the 8-fold layer of dry gauze, mackintosh, &c. A perfectly occlusive dressing is difficult to apply to the abdominal surface, especially when the wound approaches near the pubes. At this point and at the groins special care must be taken, and plenty of salicylic wool used.

The bandages, dipped before application in carbolic solution, should be applied from the lower ribs down to the groins and thighs, around which they are applied firmly in the form of a double spica. The firm compression thus realised proves of great advantage, and forms a very efficient safeguard against the entrance of air at the lower part of the dressing.

The salicylic wool applied at the epigastrium, groins, and symphysis, acts as a filter, fills up irregularities, and exercises

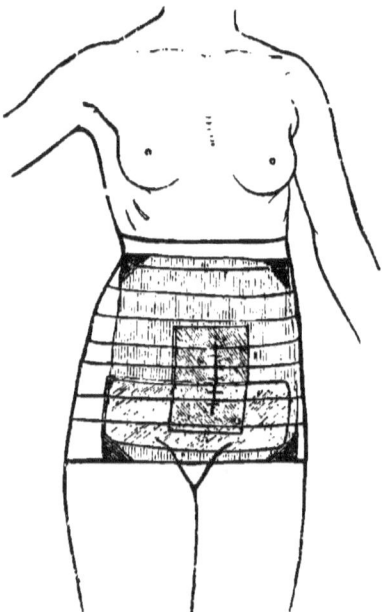

FIG. 40. Antiseptic dressing after ovariotomy.

a beneficial degree of pressure. Flannel bandages are applied over all. The second dressing may generally be safely postponed for seven days, after which the wound will be found healed, and the sutures may be removed.

The wound usually unites by first intention throughout. Even in fat or elderly women only one dressing is commonly required, and the patient in uncomplicated cases makes a painless and perfectly uneventful recovery.

In a discussion at the Medico-Chirurgical Society, on February 10, 1880, Mr. S. Wells stated that during the last two years he could not remember seeing a single drop of pus in any of his antiseptic ovariotomies. Contrary also to what one might *a priori* suppose, carbolic poisoning is rare, and when it does occur, is trifling in degree, except after very severe, long-protracted operations.

The number of spectators at an operation should be limited, for they spoil the atmosphere of the room; and any person recently in contact with putrefying matter, or with septic or zymotic diseases, should be excluded, unless after ablution and change of clothing.

A simple plan of applying an antiseptic dressing after ovariotomy is shown in fig. 40. It is the method which has been generally adopted at the Samaritan Hospital.

A, A pad of carbolic gauze, eight layers, moistened with lotion, is first applied without protective.

B, Similar pad of eight layers, put on dry and extending across below the first. In this direction any discharge which may occur will flow.

C, Large square dressing of eight layers of gauze with pink mackintosh between the two outer layers.

The horizontal cross-lines, fig. 40, indicate straps of adhesive plaster which should overlap one another, commencing from the bottom with two broad straps.

Any irregularity is filled up with salicylic wool outside the gauze, but under the straps, and the body is surrounded with a flannel bandage lined with calico.

The dressing is not changed for a week, but if it becomes loose a fresh strip of strapping may be applied along the lower edge of the dressing and a folded towel may be applied beneath the flannel bandage to keep all firm.

HYDATID OF THE LIVER.

In the proceedings of the German Surgical Congress of 1877 an operation planned and carried out by Volkmann for the radical cure of cases of hydatid of the liver is described by Ranke. Two successful cases are here recorded, and since then several others have been performed.

The operation consists in making a free incision through the abdominal wall with the most exact antiseptic precautions. This incision may need to be as much as seven or eight inches long, and is made over the most prominent part of the hepatic tumour, usually in a direction parallel to the ribs. After division of the peritoneum the surface of the liver bulges into the opening, the margins of the wound in the abdominal parietes are now drawn apart, and it is packed with loose moistened carbolic gauze. A large dressing is then superimposed, and the patient is removed to bed. In the course of a week the patient is again brought into the theatre, and on removal of the dressing the edges of the wound are found firmly adherent to the surface of the liver, a portion of which now forms the floor of the wound.

After making sure that these adhesions are complete in the whole circumference a free incision is made through the liver-substance into the cyst, and its contents are evacuated. The cavity may then, if it be wished, be washed out with salicylic lotion. Adequate drainage is provided by the introduction of a large-sized tube, and the wound is again dressed. The opening, already somewhat contracted before the hepatic incision, now rapidly lessens in size, the cyst-cavity diminishes, and meanwhile the amount of serous discharge is not very large. In the course of from three to five weeks the patient is cured, the small cysts coming away with the discharge during that period.

The second stage of the operation does not necessitate the administration of an anæsthetic, as incision of the liver-substance does not appear to be attended with much pain.

With regard to the choice of subjects for this operation it may be recommended in those cases in which puncture has failed in producing a cure; and also in those cases in which suppuration of the sac has taken place—in the latter event the double operation will sometimes not be needed, as adhesions between the liver and the abdominal wall have often already taken place.

In Volkmann's cases the operation was not followed either by local or constitutional disturbance of any moment, while the cure proved rapid and complete.

HEAD INJURIES.

The successful treatment of severe injury to the head is greatly aided by the employment of strict antiseptic precautions. The brain-substance seems tolerant to an extraordinary degree. Strong carbolic solutions, the 5-per-cent. watery, and the 10 or even 20-per-cent. spirituous, solution may be applied to it without producing any ill effect; while antiseptic occlusion seems to avert most completely those disastrous inflammations of the brain and its coverings which have so frequently caused a fatal termination after injury of the skull and its contents.

The so-called prophylactic trephining—by which I understand the removal or elevation of depressed or comminuted portions of skull in the absence of any symptoms indicative of brain-injury—will be more frequently resorted to, and with a greater advantage to the patient than was previously possible. The evil consequences, both immediate and remote, of sharp splinters of bone pressing on the dura mater and brain are universally recognised and admitted; .and, in cases where such a condition may be surmised to exist, this, and not the presence or absence of brain-symptoms, will be held as the proper indication for interference. Nor will the fact that there is no external wound be a sufficient reason in all cases for non-interference. Where a comminuted and depressed fracture of the vault exists, and can be diagnosed, it may be right for the surgeon to cut through the soft coverings, for the purpose of elevating the depressed bone; and with no greater hesitation than he would enlarge an already existing wound, confident in the security that subsequent antiseptic occlusion provides against accidents due to the presence of a wound.

Primary trephining for injury has hitherto proved so fatal that many eminent surgeons have rejected it almost altogether.

Dieffenbach said he dreaded the trephining more than the injury.

Stromeyer lays it down that a military surgeon requires no trephining instruments. Indeed, he so strongly objected to the operation, that he sums up his opinion by remarking: 'Wer heutzutage noch trepanirt ist selbst auf den Kopf gefallen.' 'He who would trephine nowadays must have had his own skull cracked.'

Malgaigne speaks of the doctrine of compression, and the necessity for trephining, as a long-standing and deplorable error.

A change must now be made in the indications for the operation, and the principles of the treatment of a fracture of the skull will come to a greater extent to resemble those which guide us in cases of fracture elsewhere.

In recent scalp-wounds the first care of the surgeon must be to thoroughly cleanse them and remove all foreign

HEAD INJURIES.

matter— sometimes a difficult thing to do, as it is more or less engrained in the tissue.

A sufficient quantity of hair must be shaved away from the neighbourhood of the injury, so that a space of three or four inches all round is clear. When the wound is extensive, and the soft parts so seriously damaged that much discharge is to be expected, or when the injury is not quite recent, it is better to shave the whole scalp. This will be also necessary where the brain is injured, as the presence of hair prevents the gauze dressing lying close upon the surface, besides being a receptacle of an undesirable kind for the discharge to accumulate in. The wound is now purified with a 5-percent. solution, the partially separated portions of scalp sutured, drainage provided for, and the dressings applied so as to extend wide of the injury in every direction. Scalp wounds unite very readily even when much contused, on account of the highly vascular supply of the part, and there need be no hesitation in introducing sutures in sufficient number to bring the edges throughout into apposition. If the wound have been exposed for several hours, or there be any suspicion of decomposition, the strong spirit solution should be employed to disinfect it.

Where, as is so often the case, the periosteum has been stripped off, and the bone is exposed, no change will be necessary in the treatment. Under this management the wound heals easily and quickly. The chances of necrosis are less; and the occurrence of suppuration, or its spread in the loose subaponeurotic cellular tissue, will be reduced to a minimum.

A compound fracture of the skull must be treated on similar principles to those adopted for fracture elsewhere. Loose or splintered fragments of bone must be removed, and depressed portions elevated; the trephine, saw, chisel, the parrot bill, or Luer's forceps being employed, according to the nature of the case. All clots must be washed away, and the exposed brain or dura mater disinfected with 5-per-cent. carbolic solution, and afterwards in some cases with the strong spirit solution, then drains are inserted, and the dressings applied. Neither the presence of a drainage-tube in contact with the brain-substance nor the application of the strong carbolic solution appears to cause any irritation or unpleasant symptoms.

The cases which I venture to quote illustrate, I think, some points of importance and interest in connection with head injuries treated antiseptically.

Hueter [1] describes a case of severe compound fracture of the parietal bone occurring in a boy through a fall of twenty-five feet from a tree. He removed a large piece of depressed bone without an anæsthetic, as the patient was unconscious. The detached fragment was rhomboid in shape, 3·8 × 4·5. c.m. in size. Half a teaspoonful of crushed brain-substance followed, coming from the termination of the Fissura Rolandi. From the wound in the skull a fissure extended into the temporal fossa, and a second fissure into the frontal bone.

A large blood-extravasation was observed in the temporal region, which appeared to increase in quantity even during the operation. The superficial wound was therefore enlarged along the line of fissure to the middle of the malar bone, exposing the upper part of the greater wing of the sphenoid.

The fibres of the temporal muscle were separated by fluid blood, and blood in quantity exuded through the fissure itself.

There could be little doubt that the middle meningeal artery or one of its branches was lacerated, and Hueter proceeded to expose and secure the bleeding vessel. Several loose fragments of bone were extracted, the wound in the

[1] *Centralblatt für Chirurgie*, August 1879.

dura mater was freely enlarged downwards, all clots were removed, and the position of the bleeding point discovered, and as soon as the vessel was tied all bleeding ceased.

The wound-cavity was then washed out with 3-percent. carbolic solution, and two drains inserted, one emerging at the upper, the other at the lower, extremity of the wound, which was then united between by twelve points of suture.

The temperature on one occasion rose as high as 101° Fahrenheit, but for the greater part of the time it remained normal. The wound healed by the first intention except where the drains emerged. The upper one was removed in eight days, the lower in fourteen days.

Absolute insensibility continued for three days, during which time the boy was fed by enemata.

The first sign of returning consciousness was an effort to swallow. Aphasia persisted till the end of the third week.

The final result was a complete restoration of brain-function.

There is always some uncertainty in the diagnosis of bleeding from the meningeal artery, and there has been great difficulty in securing the vessel in the cases in which the attempt was made.

The middle meningeal artery may be most conveniently reached by trephining almost immediately above the malar bone. The best place to apply the crown of the instrument is over the anterior inferior angle of the parietal bone, at which point the main artery spreads out in branches over the skull. This corresponds externally with the crossing of a horizontal line two fingers' breadth above the zygoma, with a vertical line a thumb's breadth behind the frontal or ascending process of the malar bone.

Professor Vogt has described in detail the best method of finding the vessel.

This case Hueter states to be the first successful one on record of ligature of the middle meningeal artery for hæmorrhage; all previous attempts having failed.

But the statement is far from correct. The two cases referred to by Mr. Prescott Hewett in his article in Holmes' 'System of Surgery' both did well, and all symptoms disappeared at once.

One was operated on by Mr. Keate, in 1839, the other by Mr. Tatham in 1842. How the mistake arose is difficult to explain. In the article on injuries of the head in Pitha and Billroth's 'Handbuch der Chirurgie,' by Professor Bergmann, both these cases are expressly stated to have terminated fatally, whereas full particulars of their complete recovery are given by Mr. Hewett in his lectures delivered before the Royal College of Surgeons, in 1854, and published shortly afterwards in the 'Medical Times and Gazette.' Mr. Cock also had a successful case of trephining for extravasation from ruptured middle meningeal artery in Guy's Hospital in 1841, and Mr. Erichsen mentions one as having occurred in University College Hospital.

Socin [1] on one occasion trephined the skull in three places in search of the source of hæmorrhage, which had produced coma from blood-extravasation between the skull and dura mater, after an injury. The vessel was reached and secured, the wound washed and drained, and a perfect recovery ensued.

COMPOUND DEPRESSED FRACTURE OF PARIETAL BONE.

W. A., age twenty-five, a carpenter, was admitted under Mr. Croft's care, December 17, 1879, with a severe compound depressed fracture of the skull, accompanied by marked symptoms of compression of brain. The accident was caused

[1] *Korrespondenz-Blatt für Schweizer Arzte*, 1879, p. 17.

HEAD INJURIES.

by a piece of metal falling on the head, from a height of forty feet. The depressed surface occupied an area equal in size to a penny, over the posterior superior angle of the left parietal bone.

After shaving the scalp and disinfecting the external wound, trephining was performed under the spray, the depressed bone being elevated, and the fragments removed. The

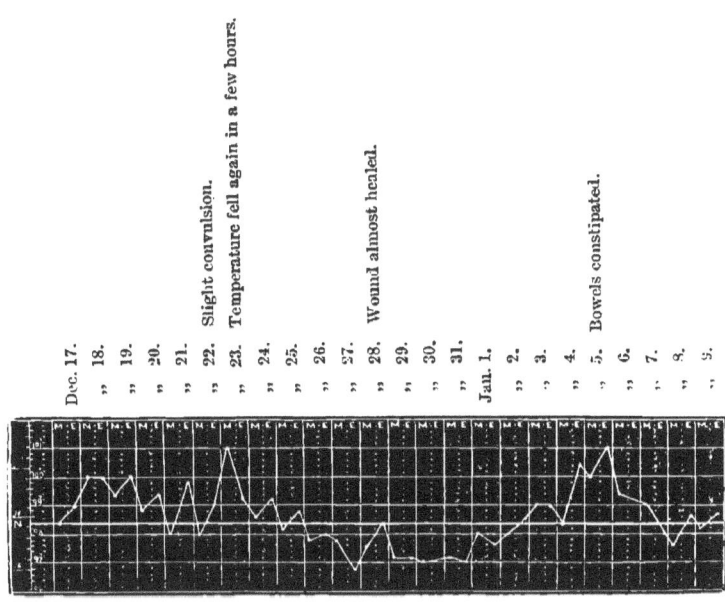

Pulse and respiration normal throughout.

Fig. 41.—Temperature chart in the case of W. A.

brain-substance was now exposed and found to be extensively injured to some depth.

The cavity thus produced was then thoroughly disinfected, first by a 1-in-20 solution, and afterwards by strong spirit solution of carbolic 1 in 10, much brain-substance coming away during the washing.

The edges of the large external wound having been carefully adjusted and united by catgut sutures, a drain was inserted into the cerebral wound, and antiseptic gauze applied in ample quantity outside.

The dressing was renewed the following day, when some blood and brain-substance were found in it.

On the sixth day the dressing was again renewed on account of some convulsive twitching of the left side of the face, and of the right hand. The wound looked perfectly healthy. In the dressing there was only a little serous discharge. The drainage-tube was removed; the catgut sutures, causing no irritation, were left.

In twelve days the wound was healed, a linear cicatrix only remaining, except at the point of junction of the three lines of incision.

During the whole time there was no purulent discharge.

Six days after admission, the temperature previously not much above normal, rose to 101, but fell again in a few hours.

A few days later a similar rise took place, which disappeared with the opening of the previously constipated bowels. The pulse and respiration were normal throughout. The patient's recovery was uninterrupted and complete.

This severe injury to the brain-substance and its coverings ran a perfectly aseptic course. The temperature chart here given, fig. 41, illustrates this very perfectly.

There was no suppuration or necrosis, no brain inflammation or alarming symptom of any kind, and the patient has been restored to health with his nervous functions almost unimpaired. The right arm is not yet· so strong as it should be normally. He is able to speak and write fairly well, though at first he was quite aphasic and of course agraphic.

TRAUMATIC LESION OF LEFT HEMISPHERE, RIGHT BRACHIO-FACIAL PARALYSIS, WITH APHASIA.

I have elsewhere published[1] an account of a similarly successful case of compound depressed fracture of the skull caused by a brick falling from a height of thirty feet on the

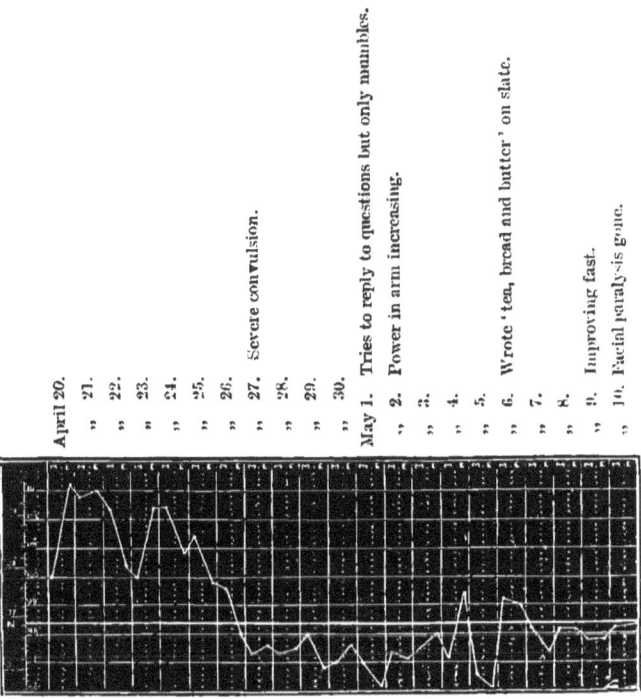

Fig. 42.—Temperature chart in the case of T. S.

left parietal bone, near its centre. Many small pieces of bone (fig. 43), a portion of cap-lining, and a number of hairs were removed from the brain-substance, which was pulpified at the seat of injury for a depth of at least three-quarters of an inch. The wound was thoroughly purified with five-per-cent. carbolic solution, two drainage-tubes inserted, the scalp wound

[1] *Brain*, vol. i. 1879.

united, and carbolic gauze dressings applied. The patient, T. S., had serious symptoms indicative of considerable brain-injury, partial paralysis of the right arm and right side of face, and complete aphasia. He had a severe convulsion on the eighth day. Nevertheless the wound united by primary union. There was no suppuration. For fourteen days he continued speechless, though conscious and intelligent. The report records that two months after the injury he appeared to be quite well in all respects, and he has since remained well. The temperature chart, fig. 42, is interesting. Within two hours of the injury the recorded temperature is 103·4, which was certainly not due to any septic change.

Fig. 43.—Comminuted portions of skull and piece of cap-lining removed in the case of T. S.

On the following day and subsequently it became gradually lower, reaching the normal standard on the seventh day, above which it practically never again rose, not even during the fit of convulsions which occurred.

The drawing, fig. 43, will give an idea of the degree of comminution. The amount of brain-injury such extensive bone-damage was likely to produce may be imagined.

During the treatment the man was evidently distressed by his inability to speak, but he did not appear to suffer in any other way.

Professor Ferrier, in a note upon the case, points out that it affords a further proof of the opinion that the occurrence of aphasia, with lesion of the left hemisphere, is not merely a coincidence, to be explained by any greater frequency of lesion of that hemisphere as compared with the right, and further, that paralysis of cortical origin is, as was witnessed in this instance, unaccompanied by loss of sensibility.

I may cite another case of a somewhat different character. W. Prior, age thirty-two, was admitted with a severe compound depressed fracture of the left frontal bone. The accident was caused by a scaffold plank of wood, three inches thick, and many feet in length, falling end-on upon the

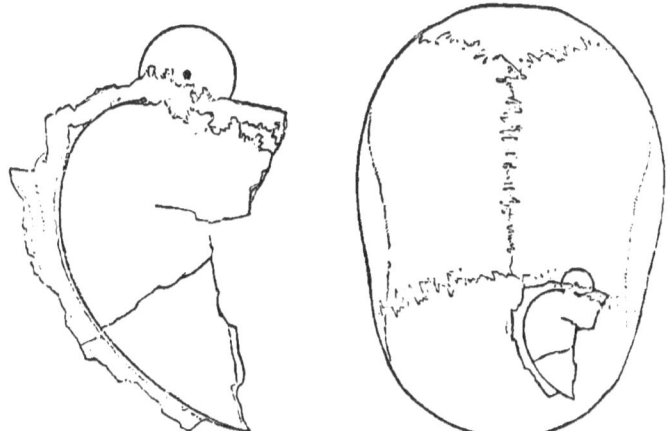

FIG. 44.—Represents the pieces of skull removed, three-fourths their natural dimensions, and the piece of bone trephined from the anterior margin of the parietal. Some smaller pieces of bone were lost. The position of the fracture on the skull is also shown. Case of W. Prior.

man's head, from a height of five-and-twenty feet. There were two small contused wounds of the scalp. The brain-pulsation was communicated to the fluid in the wound.

The skull was depressed half-an-inch at the central part of the fracture, which was comminuted. There were no symptoms of brain-injury of any kind.

Under ether, for the man was perfectly conscious, and with the carbolic spray playing on the head, I freely enlarged the external wound, forwards, backwards, and across the sagittal suture, and exposed a very extensively fissured fracture with depression occupying a large area of the bone, corresponding to the left side and upper part of the frontal bone, and causing a diastasis of the coronal suture. A fissure extended forwards towards the supra-orbital region, another to the sagittal suture, and a third downwards and forwards into the temporal fossa. The margin of the parietal bone at the suture overlapped the frontal. The bleeding was very copious, so much so as to suggest injury to the longitudinal sinus, or to a considerable artery.

I applied the trephine to the margin of the parietal bone and was then able to remove two large fragments of depressed completely detached bone, and several small pieces, beneath which, although only two hours had elapsed since the receipt of the injury, a large quantity of blood-clot had accumulated. I was unable completely to replace the external part of the frontal bone, which I was thus compelled to leave about two lines below the level of the parietal. The bleeding from the vessels of the dura mater was very free. It came from numberless points. But no important vessel seemed to bleed; and the membrane was untorn. I now removed all blood-clot from the surface, and from beneath the edges of the bone, where it was half-an-inch thick in places, washed the wound with five-per-cent. carbolic solution, and afterwards with a camel's-hair brush thoroughly disinfected every part with spirit solution 1 in 10, passing the brush an inch or more beneath the bone where the dura mater had been detached by the extravasation.

The margins of the extensive skin-wound were now sutured with gut, three drainage-tubes were inserted, each one being fastened to the skin by a gut stitch, and such an antiseptic dressing was applied as is shown in fig. 45.

HEAD INJURIES.

Next day the dressing had to be removed on account of very copious bleeding, which had saturated the gauze.

The following day the dressing was again changed for the same reason.

On the fourth day two drains were removed; the wound appeared healed by first intention, except at the drainage openings. The gut sutures were not removed; those securing the drains gave way by the time their object had been served.

The patient from first to last had not the smallest—the most trivial—symptom attributable to this extensive injury.

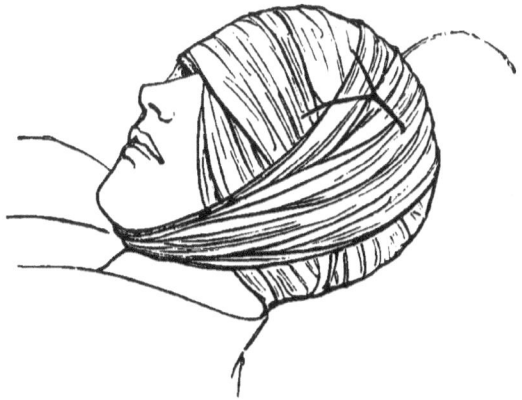

FIG. 45.—Shows the external appearance of the dressing applied in this case, and the position of the wound made in the scalp.

His voice was strong, his intelligence perfect, his bodily functions normal, except the pulse, which was slow. The temperature the evening of the accident reached 100°; next day it became normal and never afterwards rose again. His only complaint was that his bed was somewhat hard, and his appetite too good for the diet he was kept upon.

There can be no reasonable doubt that, had the depressed bone not been removed, the serious hæmorrhage would have continued, and, no sufficient outlet being provided, the blood would have accumulated between the skull and dura mater, and probably produced unpleasant symptoms. Under ordinary circumstances the loose bone-fragments if not removed would have become necrosed, and if removed the exposure of so large a surface of dura mater, without antiseptic protection, would have been dangerous to the life of the patient. Incomparably less so, no doubt, than if the brain-substance had been exposed, but still dangerous, for inflammation occurring in the wound would involve the dura mater, and only too probably extend to the crushed brain-substance lying immediately beneath. For we may assume, I think, that the violent smashing in of the skull caused by the blow could not have taken place without seriously damaging the brain-substance at the same time.

Ligature of Blood-Vessels.

The catgut ligature offers great advantages in operations upon the blood-vessels, whether for tying the arteries and veins in a stump after amputation, or still more so in the ligature of a trunk in its continuity. The advantage here is notable. The external wound may be closed, the loop of the ligature whose ends are cut short becomes infiltrated and surrounded with plastic matter, and is finally amalgamated with the external coat of the artery. The death of a portion of the external coat does not take place, nor does it ulcerate through as it previously did, before the silk ligature with its end hanging out, could be removed from the wound. Obliteration of the artery will be more secure as well as more rapid, and the danger of secondary hæmorrhage is diminished, especially in those cases where a collateral branch is given off close to the point ligatured.

Mr. Lister[1] some ten years ago published his early clinical and experimental observations on the behaviour of animal

---

[1] *Lancet*, April 3, 1869.

ligatures applied to vessels in their continuity, and on the process of their organisation.

Fig. 46, copied from Professor Lister's paper, illustrates the manner in which a ligature of this kind which had been applied to the carotid artery of a calf with antiseptic precautions became organised. The animal was killed thirty days afterwards, and the ligature was then found to be fused with the external coat, and its substance invaded and replaced by fibroplastic matter, consisting mainly of long multinucleated cells resembling fibres. Between the ligature and

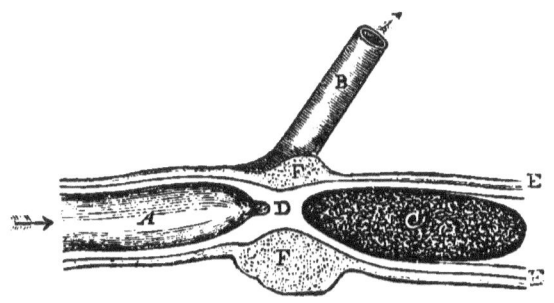

Fig. 46.—Represents the process of obliteration of an artery and the organisation of an animal ligature applied to it.

A proximal portion, B collateral branch, C coagulum, D thickened adherent internal coat, E external coat, F fibroplastic material.

the heart no coagulum had formed because of the origin of a large branch immediately above the part tied. The internal coat had become thickened, and its surfaces had coalesced so as to form a *cul de sac*, well capable of resisting the pressure of the arterial current. From these experiments it would also appear that the division of the external and middle coats by the constricting ligature is not essential, and that the coagulum in the vessel takes little or no part in the process of its obliteration. On the proximal side there is no coagulum. None could form on account of the current through the large collateral branch given off just at this point, but this has not interfered with the perfect closure of the tube.

Varicose veins may be excised, and the divided ends tied with catgut with a safety which no previous operation on veins ever attained. I have several times practised an operation of this kind with relief to the patient, and without a trace of local or general disturbance of any kind.

EMPYEMA.

Abscess in the pleural cavity is difficult to treat antiseptically, because of the mobility of the parts, and chiefly because the air is drawn into the chest with each inspiration. The spray, therefore, must be in perfect working order.

The surface of the chest on the side involved should be washed, before making the incision, with 1-in-20 solution, which will soak into the epidermis and skin follicles, and neutralise the ill effects of any dirt in the skin.

The hands and the instruments being carbolised, an incision is made into the pleural cavity at the most suitable point, under a thoroughly efficient spray. This is an absolute essential, more here than in most cases, because when the opening is made, each inspiration will suck a quantity of air into the pleural cavity, and care must be taken that no air, except that mingled with the spray, gains admission. By so doing, the risks of putrefactive change will be guarded against, and the discharge soon changes from pus to an odourless serum.

The opening made, it is a question whether or not the pleural cavity should be washed out. To do so with carbolic solution, especially in children, is to run great danger of poisoning. A salicylic lotion has no such disadvantage, but it is not so certain in its antiseptic power.

If the operation have been aseptic throughout, with

no previous external opening or communication with the bronchial tubes, and the fluid in the pleura is sweet, syringing is not required in the first instance, and it is even more undesirable to practise it at any of the subsequent dressings.

A drainage-tube must now be introduced—for the first few days the ordinary rubber-tube suffices. Extra precautions, however, must be taken to prevent its being sucked into

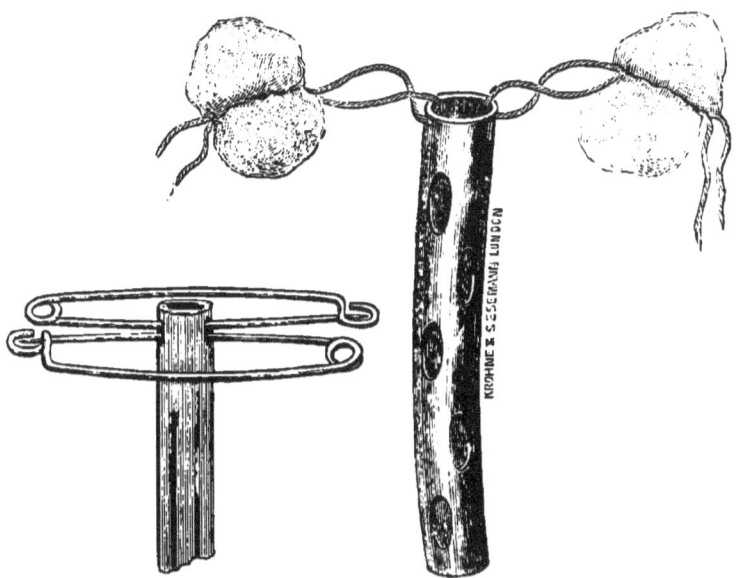

FIG. 47.—Drainage-tube, with balls of gauze attached to prevent its slipping into the pleural cavity. A similar protection is afforded by a couple of safety pins introduced as in figure.

the pleural cavity, and this may be easily accomplished by fastening to each of the attached silk threads a small ball of antiseptic gauze, so that, if the tube perchance slip in, it can be readily recovered again.

At a later period, should drainage be still required, the rubber-tube will. no longer suffice. The collapse of the chest wall diminishes the interval between the ribs, so that the tube is compressed and the flow of fluid arrested.

When empyema was treated by very protracted drainage, as it always had to be when antiseptic precautions were not adopted, it was recommended to excise a portion of rib or of two adjacent ribs, to permit a sufficiently free outlet for the discharge, and this plan has been adopted abroad as an ordinary method of practice. The substitution of a silver tube for the rubber one will often obviate this necessity; the tube must be introduced before the contraction has taken place. One made of pewter will answer all purposes.

In a case of empyema under the care of Dr. Ord, in which an external fistula had formed, and was discharging daily from 16 to 20 ounces of pus for a period of eighteen months, I recently excised two inches of one of the lower ribs. The previous opening was insufficient, being small and valvular. The fluid in the chest was not decomposed. A quart or more of pus escaped, and air now rushed in and out with each respiratory movement. The operation was done under the spray, and the chest-cavity washed out with warm salicylic solution.

There was immediately a notable diminution in the quantity of pus. In less than a week it was just enough to stain the dressing. The temperature fell, and the patient improved rapidly in all respects. The man, however, is still under treatment, and this must needs be somewhat long-continued, as the lung was completely collapsed and bound down by strong adhesions.[1]

AMPUTATION OF THE MAMMA.

The wound produced by this operation requires careful dressing, especially where glands have been removed from

---

[1] Since the text was written this patient has made to all appearance a very complete recovery. Two months have elapsed since the operation, the wound has healed, and he has increased two stones in weight.

the axilla. The gauze should envelop half the arm, and extend up to the neck, and an elastic band should be applied round the margins, so as to control the dressing during the movements of the neck and thorax. Drains are never to be introduced the whole length of the wound, but only at the extremities.

Especial care should be taken to drain the axilla. A

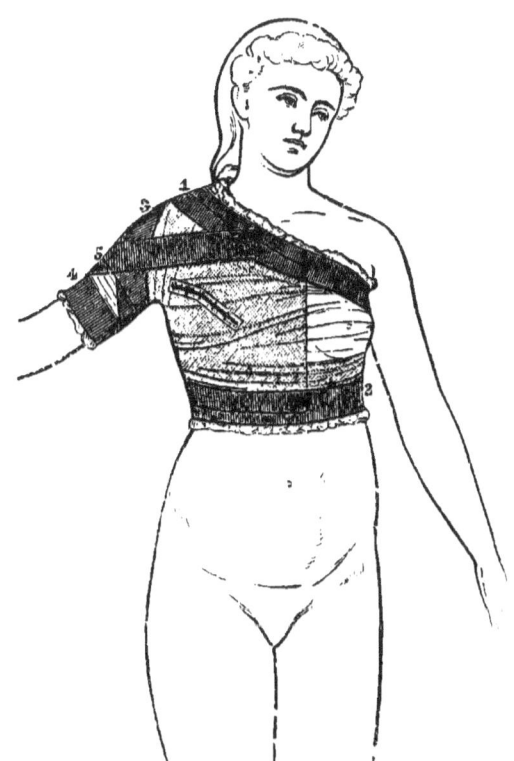

FIG. 48.—Antiseptic dressing as applied after removal of the breast.

counter-opening through its posterior wall may be often made and a drain passed straight out through it with advantage; a drain should be inserted at the inner angle of the wound. A drain here will act well when the patient is directed to lie upon the sound side, the fluids not having the tendency they might otherwise have to gravitate into the loose cellular tissue of the axilla. Where there has been any considerable sacrifice of skin the button suture is of great advantage in bringing the edges of the wound together.

Fig. 48 represents the arrangement of an antiseptic dressing in a case of removal of the breast. The outline of the layer of moist gauze and bandage first applied is indicated, as also the position of the incision. The dry eightfold dressing would be similar in shape to the moist layer, but larger.

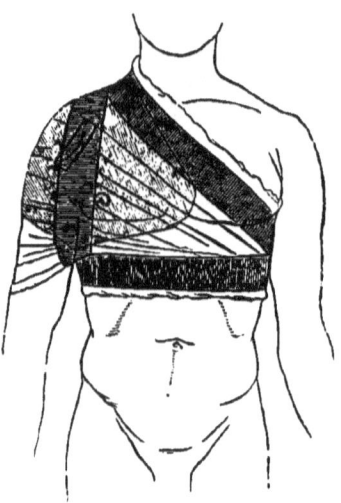

Fig. 49.—Antiseptic dressing for excision of the shoulder-joint.

The method of application of the elastic bandage is also represented. It is first applied over the shoulder on the affected side, then passed along the upper margin of the dressing in front (No. 1), and underneath the opposite arm; then passing obliquely downwards it encircles the waist at the lower border of the dressing (No. 2), and afterwards runs obliquely upwards across the back to the starting point, where it takes a turn round the arm at the border of the dressing (No. 4), and finally terminates (No. 5) by being attached to the first applied portion.

I think it may be also useful to illustrate the character of the somewhat similar dressing suitable for an excision of the shoulder-joint, of which I have had two cases recently under treatment. The woodcut (fig. 49), drawn from the patient shows the position of the wound, the form of the internal dressing, and the manner in which the elastic bandage is applied to keep it firmly in position.

ATRESIA VAGINÆ ET HYMENALIS.

To illustrate the manner in which the antiseptic method may be employed in operations in the vagina I may mention the following example.

A girl of fourteen was admitted under my care with retention of urine, for which no cause could be assigned. On examination the vagina just within the lesser labiæ was found to be occluded by a thick membrane, on which vessels of a certain size ramified.

It bulged forward, was tense and fluctuating, and a large swelling could be detected above the pubes, the dulness when the bladder was ascertained to be empty reaching close to the umbilicus; the rectum was also felt to be compressed by the projection backwards of its anterior wall. The diagnosis was not doubtful.

I operated with antiseptic precautions, made a crucial incision in the occluding septum; 35 ounces of dark treacle-like fluid escaped. The vagina was then gently washed out with 1-in-40 tepid solution of carbolic, and afterwards plugged with balls of salicylic wool, the deeper ones being first dipped in strong carbolic solution—probably carbolic glycerine, 1-in-20, would be preferable—and this was used at subsequent dressings. Outside a large mass of salicylic wool was fastened with a bandage, and a winged catheter left in the bladder.

This proceeding was repeated twice a day and under the spray, for the first three days, the object being to allow the greatly distended uterus full opportunity to contract. Some bloody fluid escaped for the first two days; but by the fifth day all discharge had ceased and the girl could pass water without assistance. The antiseptic tamponade was then omitted, and tepid carbolic injections continued twice daily. After eight days a digital examination proved everything to

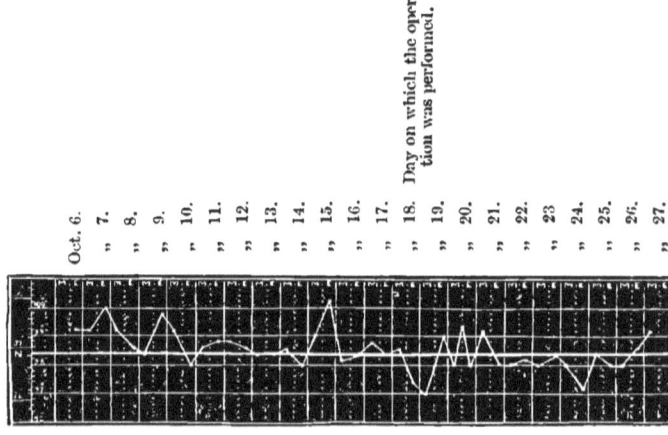

FIG. 50.—Aseptic temperature chart in a case of imperforate hymen.

be normal, and the girl got up two days later. She has since menstruated naturally. There was scarcely a rise of temperature after the operation (fig. 50), and no trouble of any kind.

Such cases have hitherto caused great anxiety to the surgeon from the rapid decomposition so liable to occur on opening the accumulation. Many fatal results have happened. If, however, decomposition be prevented during the first few days, and the uterus permitted to resume its normal size, the subsequent risks become very slight indeed. It would be easy to modify the tampon-dressing by combining drainage with it. A gum catheter might have an umbrella-shaped piece of protective silk fastened near its termination, and after introduction it could be packed all round with carbolic gauze or salicylic wool, and then, the catheter being cut to a proper length, an external dressing of gauze is applied. Or continuous irrigation might be resorted to by means of a double-channelled catheter, one end of which is connected with the irrigator, and the other with a tube to conduct the fluid after use into a receptacle near the bed. The bowels ought to be kept constipated for a few days, and the water either drawn off periodically under the spray, or, when practicable, a permanent catheter should be left in to obviate the frequent removal of the dressing.

The operation must be done with every antiseptic detail thoroughly carried out. All the folds of mucous membrane, and the external surface should be purified with the strong carbolic solution. A free incision may then be safely made while the carbolic spray is playing all the time on the part, and the dressing is completed as before described. The external dressings should be large and frequently renewed during the first few days, as the amount of discharge is often very considerable.

THE USE OF THE CATHETER.

No more distressing condition exists than that produced when symptoms of bladder-irritation, or chronic vesical inflammation, become aggravated by putrescence of the urine.

It had long been observed that in certain diseased states the use of instruments was ill tolerated, and prone to be followed by cystitis, and that the urine easily became changed from a neutral to a more or less decomposed fœtid state.

In chronic enlargement of the prostate, in stone and in stricture, decomposition of the urine usually only occurs after exposure to the influence of the air, subsequent to being voided ; but in cases demanding frequent catheterisation, it may occur within the bladder, and foul, more or less stinking, urine will be passed. That this is often due to the introduction of septic material from without, by means of a catheter, is scarcely to be doubted. The healthy urethra, which, except at periodical intervals, is a collapsed tube, does not allow the passage of bacteria along it, and the healthy bladder or urine will resist their influence. Normal urine in contact with sound mucous membrane will not easily putrefy, but urine in which mucus and blood or pus is mixed is very prone to decompose, and an impure catheter introduced under such conditions, or a catheter tied into the bladder, will almost certainly excite putrefactive changes. Not unfrequently one observes a sudden aggravation of the symptoms in chronic cases of bladder-mischief, where the catheter has to be constantly used, and the urine—which before was not in the least offensive—quickly becomes so, and is found teeming with bacteria.

The female is more liable to cystitis than the male, the short urethra being very probably the cause.

A permanent catheter should be changed every second day. An impure catheter ought never to be used. Catheters should be scrupulously clean. They should be first washed with 1-in-20 solution, and then oiled with carbolic oil of the same strength, before they are introduced.

THE TREATMENT OF HYDROCELE BY INCISION THROUGH THE
SCROTUM AND ANTISEPTIC DRESSING.

I have had but little personal experience of this method of treatment for the radical cure of hydrocele.

The ordinary plan by the injection of iodine is certainly painful, sometimes followed by severe orchitis, and also occasionally by relapse, necessitating the repetition of the process.

The operation Volkmann would substitute is thus described by Genzmer.[1]

The patient having been chloroformed, the lower part of the abdomen, external genitals, perineum, region of anus, and upper third of thighs, are washed with soap and water; then the hair is shaved, especially that on the scrotum, which is the most difficult to remove. The parts are afterwards disinfected with 1-in-20 carbolic solution.

The surface of the tumour having been made tense under the spray, the tissues are cut layer by layer, by an incision extending from the upper extremity of the swelling vertically downwards for three or four inches. As soon as the tunica vaginalis is opened, two fingers of the left hand are introduced into the cavity, and with a scissors, or blunt-pointed bistoury, the whole thickness of the anterior scrotal wall is at once divided to the full extent of the external wound.

Any bleeding points are now secured. The divided edges of the tunica vaginalis are then accurately sutured with catgut to the margins of the skin, and the cavity is washed out with 3-per-cent. carbolic solution.

Meanwhile the scrotum has contracted, a drain is inserted, and the margins of the wound are drawn together by two or

[1] 'Die Hydrocele und ihre Heilung durch den Schnitt bei antiseptischer Wundbehandlung,' von Alfred Genzmer. *Sammlung klinischer Vorträge*, April 1879.

three deep silk sutures, which serve to retain the drainage-tube in position, to close in the testicle, and render the healing process more rapid.

The most difficult part of the proceeding is the subsequent dressing. There are in this region obvious difficulties in procuring perfect antiseptic occlusion and compression.

The pelvis is raised on a support, see fig. 51, and the knees lifted up by cushions, and the wound again cleansed of any blood. Masses of folded gauze are now applied to each side of the scrotum, filling up the interval between it and the thighs; another mass is applied on the perineum behind the scrotum. The root of the penis is surrounded by salicylic wool, and a layer of the eight-fold gauze, half a yard square, is then applied over all, with a slit in its upper margin for the penis to emerge by, the lower margin is folded over the perineum, and the corners lie on each thigh. A double spica bandage is applied, and antiseptic wool packed under the margins. The dressing must be well secured at its edges, and this may be accomplished by a turn of elastic bandage round the pelvis and the thighs. The surface of the perineum is the place most difficult to fasten the dressing to. To effect this a turn of bandage may be carried from the anterior superior spine on one side along the fold of the groin, across the scrotum and perineum, which passing upwards behind circles around the iliac crest on the opposite side, and then back again to the perineum over which it passes to the starting point. This forms a kind of figure of 8. The anus and ischial tuberosities are left uncovered.

The operation and subsequent application of the dressing occupy about half an hour. The only inconvenient after-result observed, was retention of urine, which occurred in eight cases out of seventy operated upon, during a period which extended over three years.

In none did scrotal phlegmon or abscess occur. The

result, almost without exception, was a primary obliteration of the cavity of the tunica vaginalis without suppuration.

The drains are removed on the third or fourth day and the silk sutures at the second dressing, which usually takes place on the sixth or seventh day, by which time the entire wound is usually healed with the exception of a small granulating strip, and a suspensory bandage lined with antiseptic cotton or boracic lint can be applied. A slight amount of swelling and tenderness of the testis and epididymis was observed in several cases. It disappeared with the cicatrisation of the wound, in from eight to fourteen days.

Several surgeons have modified the proceeding by simply making a short incision in the scrotum, washing out the interior of the tunica vaginalis, and inserting a drainage-tube.

The objections raised to this very much simpler practice are, that drainage will scarcely be so thorough as to admit of that primary adhesion taking place which occurs after the long incision made in Volkmann's operation, which, when it takes place, must of course prevent relapse. It appears, however, that relapse has been observed in a few cases even after the long incision.

It is contended that the method generally cures the patient in a week, with little or no fever or pain, while the iodine injection causes at the time and afterwards very great pain. That a diagnosis in doubtful cases is thereby made easy, and a tumour of the testicle, of which the hydrocele is a symptom, may be thus examined, and perhaps in some cases treated by immediate removal, or in others by incision.

In simple hydrocele with a thin scrotal wall, or one not altered in any considerable degree, this method appears to furnish a very certain and easy cure; but the old method of puncture, followed by injection, will probably not be superseded for some time. In cases of old-standing disease with thick walls, or in those in which injection has failed, and in cases of disease of the testicle or epididymis, we have in the incision method, with antiseptic precautions, a valuable addition to the resources of operative surgery. In ordinary cases I think the less formidable if less complete plan, by a limited incision, washing out with five-per-cent. carbolic solution and drainage, will answer the purposes of cure.

The chief risk attending this method of treatment would certainly appear to be connected with the difficulty of maintaining a perfect antiseptic occlusion.

ON THE ANTISEPTIC REMOVAL OF CANCER OF THE RECTUM.

It might seem that here, least of all places, is the antiseptic system applicable; yet a decided step has been made in this direction.

An interesting paper has been published on the subject by Volkmann;[1] he thinks a more extensive removal of the rectum is practicable with safety than was formerly considered possible, and regards the necessity for opening the peritoneum, which must always occur when the disease extends far up on the anterior wall, as no bar to operation. Formerly such cases proved fatal, without exception, while now we often have good recoveries, when primary disinfection of the wound, frequent washing out or irrigation, and good drainage have been the practice adopted.

There are three forms of the disease for which the operation may be practised. One is a circumscribed tumour of the rectal wall situated within the external sphincter, without implication of the anus. To excise it, sufficient room may be gained by a forcible dilatation of the anus, or by an incision backwards towards the coccyx. This will permit any ordinary

[1] Volkmann's *Sammlung klinischer Vorträge*, March 1878.

sized tumour to be pulled well down into view. Indeed the mucous membrane, and tumour along with it, will often prolapse through the unresisting sphincter. In all cases sufficient room must be secured for the introduction of instruments and to apply the ligatures.

To dilate forcibly, either the thumbs or two fingers of each hand, bent to a right angle, are introduced into the rectum and drawn steadily outwards till arrested by the ischial tuberosities, in which direction the dilatation should be always made. This may require to be once or twice repeated before the contractile power of the sphincters is overcome. I have never seen any untoward consequence follow this apparently rough procedure.

For larger tumours, or those situated higher up, mere dilatation is insufficient; a posterior incision towards the coccyx, and frequently an anterior one also, must be made.

The difficulty of removal of such growths will depend on their distance from the anus, their extent, and whether they be situated on the anterior or posterior wall. New growths involving the anus are generally epithelial in character. The circumscribed tumour within the rectum may be either carcinoma or sarcoma, the latter being the more rare.

In a second form the lower end of the gut is invaded by the disease more or less all round; sometimes the anus is partially involved and sometimes it is free from implication in the disease. In the former case an incision of sufficient extent is made through the skin around the anus, the new growth separated from the loose surrounding cellular tissue, the bowel drawn down, and the diseased part cut away along with such portion of the anus as may be involved. The cut edge of the intestine above is subsequently accurately attached to the margin of the skin. When the anus is not diseased, a circular portion of the continuity of the gut may be excised, leaving a wider or narrower anal portion of healthy mucous membrane connected with the anal orifice, to which the upper portion of the intestine must be subsequently sutured. Dieffenbach's method may be here adopted. It consists in splitting the anus in front and behind, and extending these incisions inwards to the seat of the disease. Then transverse cuts are made clear of the tumour below so as to join the extremities of the former incisions, and thus make two quadrilateral flaps of sound mucous membrane, each containing one-half of the anus. The diseased mass is now carefully separated as before described, and drawn downwards, the sound gut above secured by loops, the disease cut away, and the upper and lower portions of intestine accurately sutured together. Then the first made anterior and posterior incisions are closed, large drains being inserted through them to the bottom of the wound.

Where there has been no perineal incision drainage is of necessity accomplished in a different fashion, and must be combined with a most exact closure of the wound towards the cavity of the gut, in order to exclude its contents as completely as possible. A bistoury is introduced from the surface about two to three lines from the anus, till it reaches the floor of the wound, and through this channel a drainage-tube, the size of an ordinary lead pencil, is introduced, and when securely fastened the wound in the bowel is accurately closed.

A sort of blind, external fistula is thus created, through which fluids and blood escape, and antiseptic injections may be made, if need be, while no foul matter coming from the bowel can reach the surface of the wound.

Care must be taken to plan the incisions so that if possible the edges of the wound in the bowel may be drawn together transversely. To do so vertically, would be to narrow the tube so much as probably to produce a stricture.

The abundance of the mucous membrane, however, in

the rectal ampulla, admits of a considerable portion being excised without the occurrence of any marked degree of tension or subsequent contraction.

In the third class of cases the lower end of the gut and anus are both involved all round. An oval incision must first be made round the anus, which in this case has to be sacrificed. The diseased part is then gradually separated from the loose cellular tissue of the ischio-rectal fossæ and drawn downwards. If there be not room enough, the wound may be enlarged by a median incision towards the concavity of the sacrum, made, of course, outside the now partially loosened rectum, and more space may be obtained by making a perineal incision forwards. When the tumour has been isolated, a number of loops of silk are inserted all round the intestine above the disease, to enable the surgeon to pull down the gut and afterwards suture it to the skin. Further points of suture are subsequently inserted to adjust the parts with the needful accuracy. After the diseased portion has been cut away, and the cut end of gut has been accurately attached by suture to the divided edge of the skin, several drains, about the size of a quill, are introduced.

If the peritoneum be wounded, the opening in it should be sutured with catgut. A well-carbolised sponge may be pressed over the opening until the completion of the operation. If the opening be small, it may be seized in a forceps and closed with a piece of catgut applied as a ligature. It cannot be predicted beforehand whether this accident will or will not occur. In the adult the point of the forefinger introduced into the rectum will not usually quite reach the reflection of the peritoneum in front, so that anywhere within the finger's reach the diseased portion may be removed without wounding the peritoneum. But the peritoneal pouch often descends lower, especially in persons who for years have been subject to straining at stool.

The dressing is completed by applying bundles of antiseptic gauze externally, and the rectum is plugged in the following way. A sufficiently large square piece of oiled silk is invaginated into the lower end of the gut, and its interior filled, but not too tightly, with salicylic wool. This can be easily removed when another dressing is to be applied, without distress to the patient. Layers of salicylic wool, a mackintosh cover, and a 'T' bandage finish the dressing.

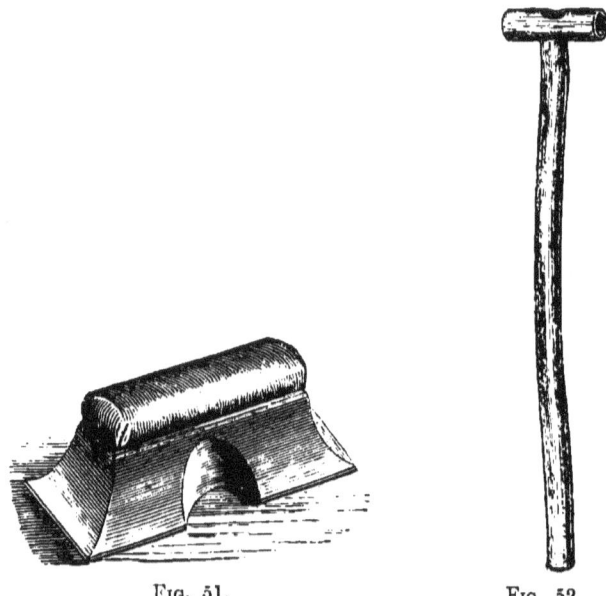

Fig. 51.   Fig. 52.

Its application is much facilitated by raising the pelvis on a Volkmann's support, shown in fig. 51.

Where the operation has been extensive, and the peritoneal cavity opened, a permanent antiseptic irrigation should be kept up for four or five days. An india-rubber drainage-tube is passed deeply into the wound, and fastened to the edge of the skin. The tubes should be perforated only near the end, and a short cross-piece attached at the deeper extremity to prevent their becoming displaced until the time comes for withdrawal (fig. 52). The success of the result greatly depends on the drainage of the peritoneum being complete.

The drainage-tube can be readily connected with the drop apparatus shown in fig. 53.

This is in turn supplied from an irrigator suspended above the bed. The india-rubber tube of the irrigator should have been rendered aseptic previously by carbolising it in 1-in-20 solution for a week.

The control of the irrigation, and the quantity of fluid used, are managed with facility by means of this simple arrangement introduced into the irrigating tube.

Salicylic solution is the best and safest fluid to use. If it flow in rapid drops through the apparatus, it suffices to keep the wound aseptic. To provide for the outflow of the fluid used during the irrigation, the buttock must be raised on the pelvic support, a cushion placed under the back, and a flat vessel arranged to receive the overflow.

After four or five days the risk of peritonitis is over, and the continuous irrigation may be then left off, and daily injections substituted. The protection cannot be absolute, but the results, on the whole, in the cases operated upon, even where the peritoneal cavity was largely opened, were highly satisfactory, as a reference to Bardenheuer's and Volkmann's essays will prove.

FIG. 53.—Schucking's glass drop tube.

The lithotomy position is the most convenient to place the patient in for all these operations. A purgative should be administered the previous day, followed by an enema. But this is not sufficient. On the operating table the lower end of the bowel must be thoroughly washed out, so that no fæcal matter may have a chance of coming into contact with the wound.

Afterwards the bowels should be kept constipated for seven or eight days. The bleeding at the time is copious, and after an extensive removal, a great number of vessels require ligature, even as many as fifty sometimes. Catgut ligatures should be employed, and a large number of clamp forceps must be always in readiness.

Any tendency to stricture which exists after the operation may often be overcome by bougies, but if a stricture form in spite of this, a division of it towards the sacrum will afford the patient very considerable relief.

Cancer of the rectum is not so malignant as in many other places, and a relapse is sometimes delayed for years. This fact offers much encouragement to perform an early eradication of the disease; besides which, it enormously relieves the sufferings of the patient at the comparatively small expense of an imperfect control over the action of the bowels.

The disease is one which affects all grades in society, but unfortunately too often the patient either does not seek relief till too late, or is allowed to drift beyond the period when a local complete removal is possible, and all that can be done is to provide an artificial opening in the groin or loin for the escape of the fæces.

Where there has been a long history of rectal trouble of some kind, piles, constipation, or catarrh, and when to this is added a discharge of blood or matter, no time should be lost in making an examination of the interior of the bowel.

Unfortunately in too many instances the spread of the disease, the extensive infiltration of neighbouring parts, the presence of multiple fistulæ, and general condition of the patient, combine to render any operation impracticable.

REMOVAL OF A TUMOUR FROM THE WALL OF THE CHEST.

As an example of a remarkable operation which would scarcely have been undertaken without the confidence and security which antiseptic precautions afford, the illustrations

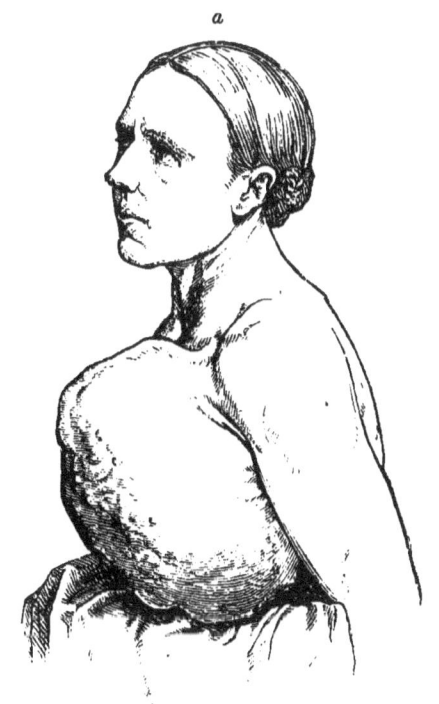

Fig. 54.—Large enchondroma growing from the ribs.

here given, drawn from photographs, afford a striking instance. The particulars of the case, which was operated upon by Professor Fischer of Breslau, were communicated to the eighth Congress of German surgeons.[1]

---

[1] 'Vorstellung eines Falles von Ausgedehnter Resection mehrerer Rippen wegen eines Chondroms,' von Dr. Kolaczek, in Breslau. Langenbeck's *Archiv.* Band xxiv.

The patient was a woman forty-eight years old, and the tumour commenced to grow some four years before from the fourth rib. It extended from the clavicle to the lower ribs, and projected outwards twelve centimètres, or about five inches. The tumour was removed with antiseptic precautions, but without spray, and with it a large portion of the fourth,

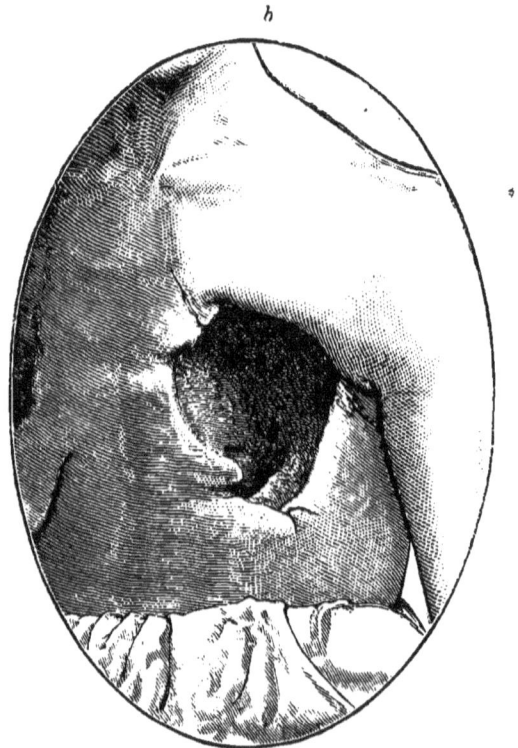

Fig. 55.—Appearance of chest-wall twelve months after operation.

fifth, sixth, and seventh ribs, with the adherent pleura, making an opening as big as a child's head, in which the lung, pericardium, and diaphragm lay exposed.

After removal of the tumour the edges of the wound were drawn together and the cavity closed over by the loose skin. Drainage-tubes were introduced by means of which it was subsequently washed out with salicylic solution. A rise of temperature which took place for four days is ascribed to a moderate amount of suppurative bronchitis; for at the first dressing, on the third day, the local reaction in the wound was slight, the skin was drawn inwards, and glued by adhesion to the pericardium and lung.

In four weeks the patient returned home. The depression left in the chest-wall is but little changed after a year's interval, when the photograph of the result was taken (fig. 55); its depth is nine centimètres (about four inches), by twelve centimètres long, and ten centimètres wide. All the heart-movements, even the pulsation of the coronary arteries, can be plainly seen.

Fig. 54 shows the lateral aspect of the tumour, which consisted of hyaline cartilage, with bone-deposits scattered through it.

Fig. 55 illustrates the cavity left after the operation, the less dark inner part corresponding to the prominence caused by the heart.

UNION BY SUTURE OF DIVIDED NERVES AND TENDONS IN OPEN WOUNDS.

Some brilliant cases of suture of divided nerves are on record in which the aseptic condition of the wound undoubtedly favoured the speedier repair of the nerve substance.

A remarkably interesting case of nerve-suture, with complete restoration of function, is recorded by Langenbeck, in the 'Berlin Med. Wochenschr.,' for Feb. 1880.

The musculo-spiral nerve had been divided, in an incised wound, where it winds round the outer side of the arm beneath the triceps muscle. Paralysis of the extensors and supinators followed, and at the time of operation, eleven weeks later, the fingers were completely flexed, the hand was extremely pronated, and the thumb adducted. The nerve-ends were exposed by an incision and found to be separated by a distance of 2 centimètres. The ends were cut off and the two portions were, with great difficulty, united by suture. These ends, when examined microscopically, exhibited no nervous elements. The wound was then closed and drained, and in five days was completely healed. On the fourteenth day slight power of extension had returned in the hand and fingers, while on the nineteenth the muscles reacted well to electricity. I saw this patient in April, there was no apparent difference between the function of the two arms. The completeness and rapidity of the restoration of function, generally much less perfect in degree than sensation, was very striking.

Gluck has even succeeded in dogs and rabbits in transplanting portions of the ischiatic nerve an inch long from one animal to another. The transplanted portion was united by sutures to the ends of the similarly divided ischiatic nerve of the other animal, and became presently healed in its new position. The nervous conducting power was subsequently re-established, and signs of muscular irritability became manifest, in some instances, as early as eighty hours after the operation. Under favourable conditions the antiseptic protection procured an immediate union of the wound, by which alone it is possible that so marvellous a physiological restoration could take place.

Wounds of the hand or wrist are often associated with a division of some of the many tendons in this region, and although occasionally single tendons have united when the wound has closed by first intention, union has probably never taken place after division of the flexor tendons in the common sheath, or when suppuration has occurred either in the tendon-sheaths or in the surrounding parts.

SUTURE OF TENDONS.

With the advantages which the antiseptic method affords, we may hope in many instances of severe injury to preserve the function of the hand and fingers, which, when many important tendons remain ununited after division, must, if preserved at all, be preserved only as a more or less useless stump.

To procure the best results, especially in wounds of the palm of the hand, the risk of secondary bleeding must be averted, and all immediate hæmorrhage arrested. The divided vessels should be secured at the seat of injury, and double ligatures applied in the wound itself. The assistance given by Esmarch's elastic constriction is invaluable for this purpose. The ligature of the divided vessels *in loco* gives the only adequate security against secondary hæmorrhage, which, did it arise, would spoil the chances of obtaining union.

The ends of the divided tendons must be sought for. A flexed position of the hand will assist the surgeon in reaching the retracted proximal extremities, but in some cases he will have to incise the sheath to get at them. It is not of first importance what material is employed for sutures. The essential thing is to procure union of the wound without suppuration, for if the latter take place no suture will long hold the tendon-ends together. Catgut is the preferable form, and the substance of the tendon should be taken up sufficiently far from its cut surface to prevent the suture from tearing out. The sheath, if largely opened, should also be closed by fine points of suture.

Drainage-tubes having been inserted, the external wound is closed. The operation is done under spray, the wound thoroughly disinfected, and the ordinary form of dressings applied externally. The hand should be kept flexed in a splint to relax the muscles concerned. Associated injury of the carpal or metacarpal bones or joints need not interfere with the method of treatment, or successful issue of the case.

Koffmann mentions [1] three interesting cases of this kind. The first was that of a young man who received from a heavy knife a deep wound in the palm reaching to the bone, and extending from the metacarpo-phalangeal joint of the middle finger to the unciform bone. The tendons of the flexor sublimis and profundus of the third, fourth, and fifth fingers were divided. Six small arteries required ligature. The six tendons were united by sutures. In sixteen days the wound was healed, and passive motion begun; complete active movement was soon re-established. When the patient was last seen, nine months after the accident, the only difference observed between the power of the two hands was that the terminal phalanges of the injured fingers were less movable than normal.

In another case, two of the extensor tendons were successfully sutured, and in a third three of the flexor tendons. In this last case the metacarpal bones were extensively fractured.

I may quote as a good result following the suture of tendon the following interesting case reported by Dr. Kölliker.[2] The patient, a boy of twelve years of age, had pushed his hand through a thick pane of glass. On arrival at the hospital an hour after the injury an incised wound six centimètres long was found on the dorsal aspect of the right wrist, extending obliquely from the ulnar side above to the radial side below. In the bottom of the bleeding wound lay the distal ends of the divided and widely separated extensor tendons as follows: the tendon of the extensor indicis, the tendons of the extensor communis digitorum of the second, third, and fourth fingers; the tendon from the extensor communis for the little finger, together with the extensor minimi digiti. The tendon of the extensor carpi radialis

[1] *Korrespondenzblatt für Schweizer Arzte*, 1878.
[2] *Centralblatt für Chirurgie*, Feb. 1880.

brevior was also three-fourths divided, the sheath of the extensor secundi internodii pollicis opened, and the tendon laid bare to a large extent.

In addition the wrist-joint was opened between the cuneiform bone, which was itself slightly nicked, and the interarticular fibro-cartilage.

The ends of the seven divided tendons were brought together with fine catgut sutures, silk sutures were used for the skin wound, and drainage-tubes were introduced.

After the application of an antiseptic dressing the hand was fixed, by means of a splint, in a position of extreme dorsal flexion, or, as we should call it, extreme extension.

The further progress of the case was extremely simple. With the exception of a temperature of 100·7 on the third evening, the patient remained completely free from fever. On the second day one drainage-tube was taken out, on the fifth the sutures were removed from the skin wound, and on the eighth the remaining two drainage-tubes. On the sixteenth day the patient could already move the fingers well. Three months later the position of the hand was normal, flexion and extension of the fingers were good, but the fist could not be completely closed. Abduction, adduction, as also dorsal flexion of the wrist-joint, were quite normal. The functions of the hand were efficiently performed. For some time past the patient had been able to write, to lift weights, and without doubt the hand-movements will eventually be perfect.

Dr. Pauly, of Posen, recounts [1] the case of a boy who fell from the top of a hay waggon, with the left heel right across the blade of a scythe.

A wound, ten centimètres, or four inches, long, extended transversely from one malleolus to the other, completely dividing the tendo Achillis, and opening the ankle-joint from

[1] *Centralblatt für Chirurgie*, Jan. 1878.

behind. The tuberosity of the os calcis, with the piece of tendo Achillis attached to it, about an inch in length, was completely severed from the rest of the bone, to which it remained connected by soft parts alone. Antiseptic treatment was adopted. The detached tuberosity was fastened in its place by a common nail, and while the foot was maintained in the equinus position, the divided surfaces of the Achilles tendon were united together by catgut sutures inserted alternately deeply and superficially into the substance of the tendon.

In nineteen days a complete recovery ensued, and one year afterwards the report states that the function of the limb was perfectly normal, '*Keine Spur einer Bewegungsstörung.*'

Comment cannot enhance the brilliancy of a result like this.

The subcutaneous division of the tendo Achillis is a harmless proceeding, but the history of surgery and Delpech's famous case tell us how different it becomes with an open wound treated in the ordinary fashion. Here in addition a piece of the os calcis was sliced off, and the ankle-joint opened into as well.

THE MANAGEMENT OF WOUNDS IN WHICH SEPTIC CHANGES HAVE TAKEN PLACE.

The antiseptic method attempts, but with a much smaller measure of success, to convert septic into aseptic wounds. Dry dressings are useless for this purpose. Carbolic irrigation is the surest means we possess, but the danger of carbolic acid poisoning is a great objection. Very strong solutions appear, however, to be less liable to become absorbed to a dangerous extent than weaker ones. Salicylic acid is too expensive for general use, and thymol as an antiseptic is too feeble. A solution of the hyposulphite of soda has, in Schede's hands, produced excellent results, and the acetate of alumina is also likely to prove useful.

The difficulty and uncertainty in dealing with cases of this kind depend on the non-localisation of the mischief, and the inaccessibility of some of its foci.

It has been stated that even the presence of a fistula [1] prevents the application of the antiseptic method. Though this be far from correct, it is sometimes impossible, or very difficult, to arrest the putrefactive process already started in an exposed wound. Under such circumstances the wound is swollen, inflamed, discharging more or less fœtid pus and broken-down blood-clot. Very energetic measures are re-

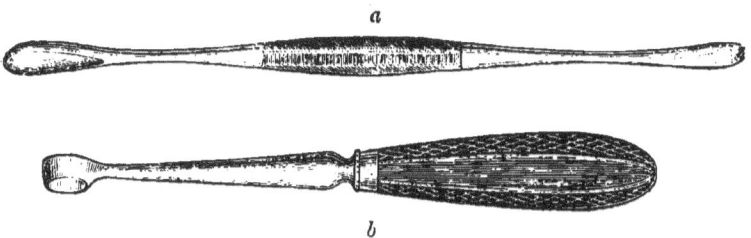

FIG. 56.—a. Sharp spoon. b. Bone gouge.

quired to check the progress of the mischief. In chronic suppurating cases, where fistulæ exist, the diseased granulation-surface must be removed, or destroyed, before the aseptic condition is realised. No more efficient or satisfactory means exist for getting rid of it than Volkmann's sharp spoon (fig. 56, a). It scrapes away all the soft inflammatory material, but does not interfere with healthy soft parts beneath.

It can also be used for the clearing out of carious cavities in bone. When the bone is hard, the instrument, fig. 56, b, answers the purpose better. The lower edge is sharp, the

[1] Letter from Professor Spence, *British Medical Journal*, Jan. 24, 1880.

tip of the forefinger rests on the opposite side, and guides and controls the instrument.

Two things must be accomplished to convert a septic into an aseptic wound. The first, most important, and most difficult, is to disinfect the wound; the second consists in affording it protection from any new infection from without. If small externally, the wound must be enlarged, so as to make all portions of it accessible, any blood-clot or disorganised tissue should be removed, and the parts thoroughly soaked with 5 per cent. carbolic solution. If any skin or tissue be infiltrated with putrid fluid, it should be cut away, or scarified deeply. The only hope of achieving success is to disinfect all parts of the wound, and the tissues surrounding it completely. In many kinds of injuries this is a difficult matter.

In all irregular wounds the recesses must be reached by the disinfecting fluid, and counter-openings made according to the rules already laid down for compound fractures. If the first attempt be not successful, the attempt must be repeated. Success follows in proportion to the completeness of the disinfection. In an open wound, under otherwise favourable circumstances, this is usually not very difficult to accomplish, but when septic changes have been going on for some time in irregular cavities, or in cases of septic phlegmon extending into the intermuscular spaces, in suppuration of the articulations, or in severe compound fractures, it is always difficult. As I have before mentioned, there can be little doubt that the tissues of some persons are more prone to putrefactive change than others, and much also depends on the circumstances and surroundings in which the individual is placed after the receipt of injury. The fearful rapidity with which decomposition is set up, and septic 'intoxication' occurs, amongst wounded soldiers crowded together—men, otherwise young and healthy—is proof of this, were proof required.

The chloride of zinc solution is an excellent disinfectant in cases already septic. It may be applied to the surface by compresses of wool or lint soaked in the fluid, and with these the wound-cavity should be filled. Smaller wounds or sinuses may be injected with a small syringe, but carbolic acid, from its volatility, is more successful and far-reaching.

As soon as the aseptic condition is arrived at, the use of strong applications is left off. The dressing of the wound is then conducted in the usual manner, taking care to apply the moist gauze direct to the surface in large quantities, and to renew it sufficiently often. Protective is not desirable, as it interferes to some extent with the escape of the secretions, and their becoming at once absorbed and disinfected by the dressing, a most necessary thing to provide for when there is any doubt as to their complete asepticity. The dressing should be changed once or twice daily, according to the state of the wound and the amount and character of the discharge.

Carbolic acid poisoning in a severe form is seldom seen, but the urine often becomes dark in colour, and troublesome vomiting is sometimes observed.

In cases of diffuse phlegmon, and in that form of suppuration which rapidly extends from the hand, up the forearm, along the sheaths of the tendons, after an injury to the fingers, in suppuration beneath the palmar or plantar aponeuroses, it must be apparent that there are great practical difficulties in disinfecting the parts involved.

In such cases we must freely make many short incisions into the inflamed tissues, and into the suppurating tendinous sheaths. The wounds thus made are afterwards washed out with 5-per-cent. carbolic solution, and numerous small drains inserted. Serum and sero-pus will drain away in large quantities, and the inflammatory tension and pain will be shortly relieved.

Hueter calls the wounds thus made button-holes. In some cases he has inserted twenty to thirty drainage-tubes, in an inflamed fore-arm, for instance. He insists that all the inter-muscular spaces must be opened and drained, and that if this be done, an aseptic condition is arrived at in two or three days. The procedure must be repeated if any fresh signs of inflammation exist; that is, further incisions must

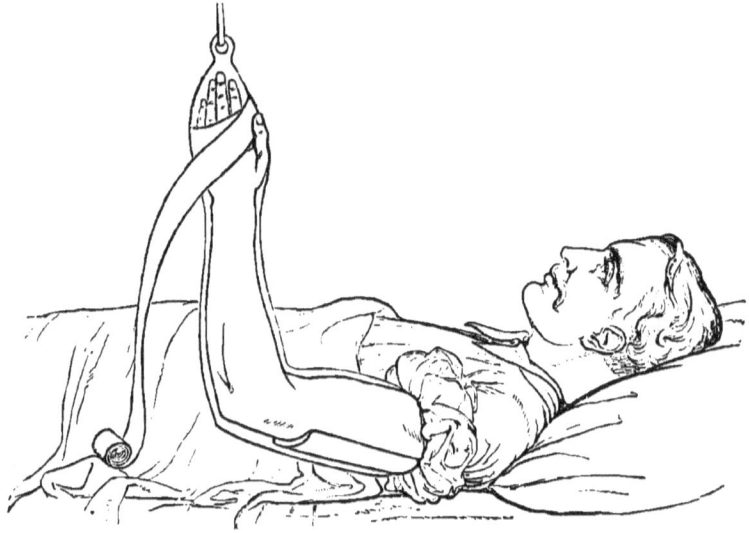

FIG. 57.—Volkmann's suspension splint. The cross-piece is made reversible, so that the same splint will answer for either arm, and for both the external and internal aspects.

be made when needed, and all the wounds washed out again with the strong solution.

König[1] has published some cases of diffuse suppuration in the tendinous sheaths of the fore-arm muscles, in which he laid open the sheath, allowed pus to escape, and then

---

[1] König, 'Antiseptisches Verfahren bei Infectiosen Eiterungen,' *Deutsche Zeitschrift für Chirurgie*, vol. x.

washed out the cavity. There was no subsequent necrosis of the tendons, and the function of the hand was restored.

Probably, on first seeing a patient affected by such an inflammation, we find the hand and fore-arm tense and swollen, dull red in hue, pitting on pressure, exceedingly painful, his general condition one of high fever or already betraying signs of septic poisoning.

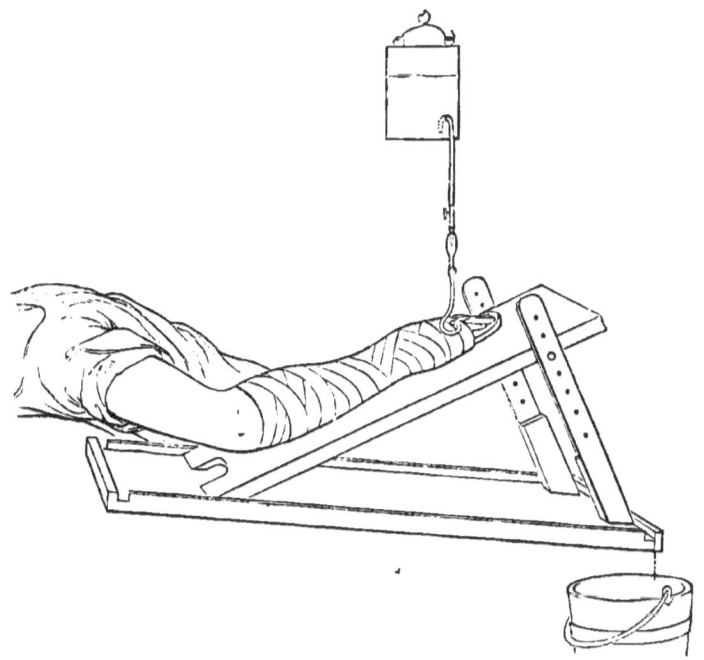

Fig. 58.—Apparatus for continuous irrigation.

The well-marked sulcus produced by the annular ligament is very characteristic of inflammation of the deeper parts in the hand and fore-arm.

As a preliminary or adjuvant to the antiseptic treatment, the arm may be suspended in the splint shown in fig. 57, and much of the inflammatory œdema will be thus removed. With the suspension continuous irrigation may be usefully combined.

The hand and arm should be washed with soap and water, and then with 5-per-cent. carbolic solution; the needful incisions made in sufficient number, under the spray, the suppurating sheaths or bursæ opened and washed out, and the hand and fore-arm enveloped in a large quantity of moist gauze, and then with the dry dressing. If a wound be already present, it must be very thoroughly disinfected. The next day, or sooner sometimes, a new dressing must be applied. If the discharge be then slight, mucous in character and inoffensive, the washing out need not be repeated, and the case will probably do well; but when it continues to be copious or ill-smelling, the freest possible disinfection should be again resorted to, and the dressings frequently changed.

In cases where the inflammation is not thus checked, continuous irrigation, or a carbolic bath, should be substituted. Some cases of acute inflammation are so rapid and intense, or have proceeded so far before treatment is commenced, that the proceedings above described afford but a slight chance of success.

Nevertheless, the practice affords the best prospect of cutting short the inflammation and the necrosis of tendons and cellular tissue so often observed, and so often liable to be followed by an acute form of septicæmia.

In cases where abscesses in the sheath of the tendons or inter-muscular spaces have already been opened, the fistulæ must be enlarged and free drainage given; the abscess-cavities are washed out, day after day, or brushed with chloride of zinc solution till a more healthy action is set up.

Good results can be achieved by carbolic irrigation, or with the salicylic solution, the arm being suspended either in the splint shown in fig. 57, or elevated by means of the arrangement shown in fig. 58. A drain may be sometimes inserted with advantage through the wound in the hand, and the nozzle of the irrigator introduced into the mouth of the drain, and the fluid then allowed to flow drop by drop till it emerges by the counter-openings in the fore-arm, placed at a lower level. When the swelling and fever diminish, and the discharge becomes sweet, the ordinary form of dressing is to be substituted.

THE USE OF ANTISEPTICS IN MILITARY SURGERY.

Anyone who has witnessed the sad sights, the accumulation of suffering, which a great battle-field presents, the neighbouring houses filled from garret to cellar with wounded, the ditches containing poor fellows who have not been picked up, lying, perhaps for two or three days, without assistance of any kind, the confusion, the insufficiency of aid of all kinds, the fewness of the surgeons, and the frequent absence of surgical supplies and ambulances from the places in which they are most wanted, will realize to the full the difficulties of applying a strict antiseptic treatment on the battle-field.

Yet a trial of this method has been made, and must again be made, though probably in some modified form, on the next opportunity.

Successful results have been obtained under very unfavourable circumstances. We must not despair of success. The practice may be simplified. Even if only a partial gain be made, this will be a benefit not merely to those who have been treated antiseptically, but an indirect advantage to the rest, by diminishing the risks incident to the accumulation of large numbers of wounded in the same place.

In field hospitals, or wherever aid to the wounded is organized, and a stricter antiseptic treatment can be carried out, it will be necessary, in order to do this successfully, for all concerned to be thoroughly instructed in the principles of the method. From the first dressing lines we must for the most part only try to deliver over the patients in as safe a condition as practicable, by means of primary antiseptic occlusion, to the places where the permanent dressings may be applied.

Esmarch's valuable papers delivered at meetings of the Congress of German surgeons, and the discussions which took place upon them, deal with many of the difficulties of the case. Gunshot wounds are inflicted on healthy men in the prime of life, and their danger often arises more from external circumstances—over-crowding, want of after-care, long transport—than from the nature of the injury itself.

The injury presents many features of what are called subcutaneous wounds. There is a small external opening which bears no proportion to the extent of the damage within, and it has been long observed that some gunshot fractures, and wounds of joints even, may heal just as simple fractures do, and the inference is that they healed because of the closure of the external wound, and the absence of suppuration depended on the non-admission of septic elements, at all events, in sufficient quantity to excite decomposition.

We now know that it is not the presence of the ball, nor the fact of the bone being splintered, which occasions inflammation and suppuration, but the entrance of septogenic matter from without, or pieces of soiled clothing carried in by the ball. The later experience of military surgeons shows that nothing is more disastrous to the possibility of an aseptic course than the 'regulation' search immediately after the receipt of the injury, and the repeated examination of the wound by the finger or instruments to discover the ball, or to diagnose the extent of the damage to the bone. It is impossible for either finger or instrument to be clean.

Formerly such examinations were necessary to determine on the propriety of amputation, but the opening of a wound to the entrance of air and septic matter destroys its chance of remaining aseptic. In any case in which a conservative treatment appears practicable, the indication would appear to be to occlude the external wound with an antiseptic tampon till such period as a more perfect dressing becomes possible. Sometimes the plan will succeed, at other times it will fail.

The relatively clean and subcutaneous character of a rifle bullet wound, its slight tendency to bleed, and the trifling amount of secretion, afford in many cases a good prospect of antiseptic occlusion being applied with success, and the wound closed to external influences for twenty-four or even forty-eight hours until circumstances admit of a more complete disinfection and dressing. And this primary treatment will very often avert the necessity for any further interference.

The extraction of fragments of bone or of the ball may be most safely postponed till a favourable opportunity, or until the symptoms declare that interference is necessary. The question comes really to be one between the preservation or the amputation of the limb, and there can be little doubt which is the preferable course when practicable.

If no fever or other symptoms declare themselves, a more permanent dressing is applied and success follows. When septic changes have set in, a careful and full examination under narcosis, with enlargement of the wound, must be made, and the future treatment will then have to be decided upon.

I think we are bound not to be discouraged, because the complete Listerian treatment cannot be applied in every case as thoroughly as we would wish, or as is practicable in civil practice.

It has now been shown that by the methodical use of an antiseptic tampon results as good as any may be obtained in war time ; that some joint injuries and compound fractures which previously were nearly always fatal, or, at all events, often demanded amputation of the limb, may be saved, and saved also in the former class of cases with a freely movable articulation.

At the first dressing places the most important duty will be the selection of fitting cases for immediate treatment, and in a future war first-rate surgeons must be selected for this task, as upon their judgment will depend the lives of many men. Those cases only can be immediately treated which most urgently demand it, more particularly the fractures and injuries to joints, large open wounds, shell smashes, or cases complicated by bleeding.

Many gunshot wounds may be safely left over, those produced by bullets with small external openings and of a valvular character, or mere flesh wounds. An antiseptic tampon will make such as these safe for many hours, perhaps permanently.

Although gunshot injuries may heal under a scab, we can never be certain that it will be so, and that no infection of the wound has occurred.

Therefore, as far as possible, the primary antiseptic occlusion should be regarded as an expedient to gain time, as being provisional to a more complete and perfect dressing which can be made later when for any reason it appears desirable, or in any case where evidence exists that inflammatory reaction has begun.

The first dressing place is not suitable for the performance of either amputations or resections, and the examination and probing of wounds there should be prohibited. A sterile examination only to be repeated at the next station or hospital can serve no useful purpose, while it does the patient a positive injury. The introduction of a blood and dust covered finger or instrument may do irreparable damage, often destroying the chances of an aseptic course.

The application of a first dressing is all that can be

attempted as a rule in the first line, and that must be, from the necessities of the situation, of a very simple kind.

Esmarch advises that every soldier shall be himself supplied with the materials necessary for a first dressing, to be carried in the breast pocket of his coat, and sufficient to afford, when applied to the wound, a possibility of its forming with the blood an aseptic scab.

In a field hospital the complete antiseptic dressing of a recent wound should be performed in the following manner.

First, carefully wash and disinfect the neighbourhood of the injury; then, under carbolic spray, dilate one or both openings and make a complete examination of the wound. If the ball has lodged, it can be extracted; if wedged in the bone, its removal is sometimes difficult, and may require the use of a chisel or gouge. Any foreign matter will, of course, be removed at the same time.

When only one opening exists, the bullet may be extracted through it, or it is perhaps more easily reached from the opposite side of the limb or body, where the counter-opening made for its extraction will also facilitate drainage. If a conservative treatment be decided upon, the wound is thoroughly washed out with strong carbolic solution, drainage-tubes must be inserted into both openings when they exist, as also into any wound-recess, through a counter-opening made when needed for the purpose.

The drainage must be especially perfect in wounds implicating joints. In the knee-joint a free incision on each side of the joint is generally required, carbolic solution must be freely injected, and made to pass by pressure into every part of the articulation.

If the wound be rendered aseptic in the first instance, all further syringing or irrigation must be studiously avoided during subsequent dressings.

In one of Kraske's cases of gunshot fracture, a coagulum 12 centimètres long became organised into cicatrix without suppuration. This could not have happened if daily syringing had been practised. The bone wound is stated to have united in this case just as easily as a subcutaneously divided tendo Achillis. In gunshot wounds of the chest, the pleural cavity, if opened, had best be washed out with warm salicylic solution. The same fluid may be likewise employed in abdominal wounds. Carbolic acid applied over so large a surface would give rise to symptoms of carbolic acid poisoning, but in cases of evident sepsis it is better to use carbolic solutions notwithstanding.

In future wars we may feel certain that in cases of penetrating abdominal wounds patients will not be left to die without an effort being made to save them. The abdomen will be laid open, the wounded intestine sutured, the cavity cleaned out, and treated subsequently as after an ovariotomy.

The washing out must be repeated over and over again in cases where any tendency to sepsis continues, and a good result may often be achieved in very unfavourable cases. The temperature will fall, and the secretions and wound become healthy.

In cases of septic phlegmon in connection with fractures, and wounds involving the intermuscular spaces, hitherto so frequent after gunshot injury, numerous comparatively small incisions should be made into the infiltrated tissues: abundance of yellow serum escapes, perhaps some abscess-cavity is opened and washed out, and great relief is afforded. The after-treatment consists of moist carbolic dressing, often renewed, with frequent washing out, or continuous irrigation with 5-per-cent. carbolic solution.

In *résumé*, the future treatment of gunshot injury upon the battle-field itself must be conducted, so far as practicable, on antiseptic principles, although at present it may be difficult to completely carry out the details of the practice.

We must endeavour, in Esmarch's words, to do the least possible harm: '*Nur nicht schaden.*' We know a large number of the subjects of severe gunshot injuries may recover after occlusion, and therefore the antiseptic tampon affords a fair prospect of a considerable success.

The packet of dressings which Esmarch would supply to every soldier (fig. 60, 1) consists of

2, The well-known triangular bandage, made of cheap unbleached cotton, and a safety pin for fastening.

FIG. 59.—Antiseptic tampon, consisting of chloride of zinc jute enclosed in gauze.

3, A gauze bandage 2 mètres long, and 11 centimètres broad, with a safety pin.

4 and 5, Two antiseptic balls of salicylic jute, of wool contained in salicylic gauze, and enclosed in a square of oiled paper; the whole makes but a small packet, 12 × 9 × 2 centimètres, which is protected from wet and external dirt by an envelope of parchment paper. The parcel which Professor Esmarch lately sent me, with its contents, is represented drawn to scale in fig. 60. Bardeleben and Münnich recommend in place of the salicylic acid wool a tampon of chloride of zinc jute, which is a more powerful and certain anti-

T

ANTISEPTIC SURGERY.

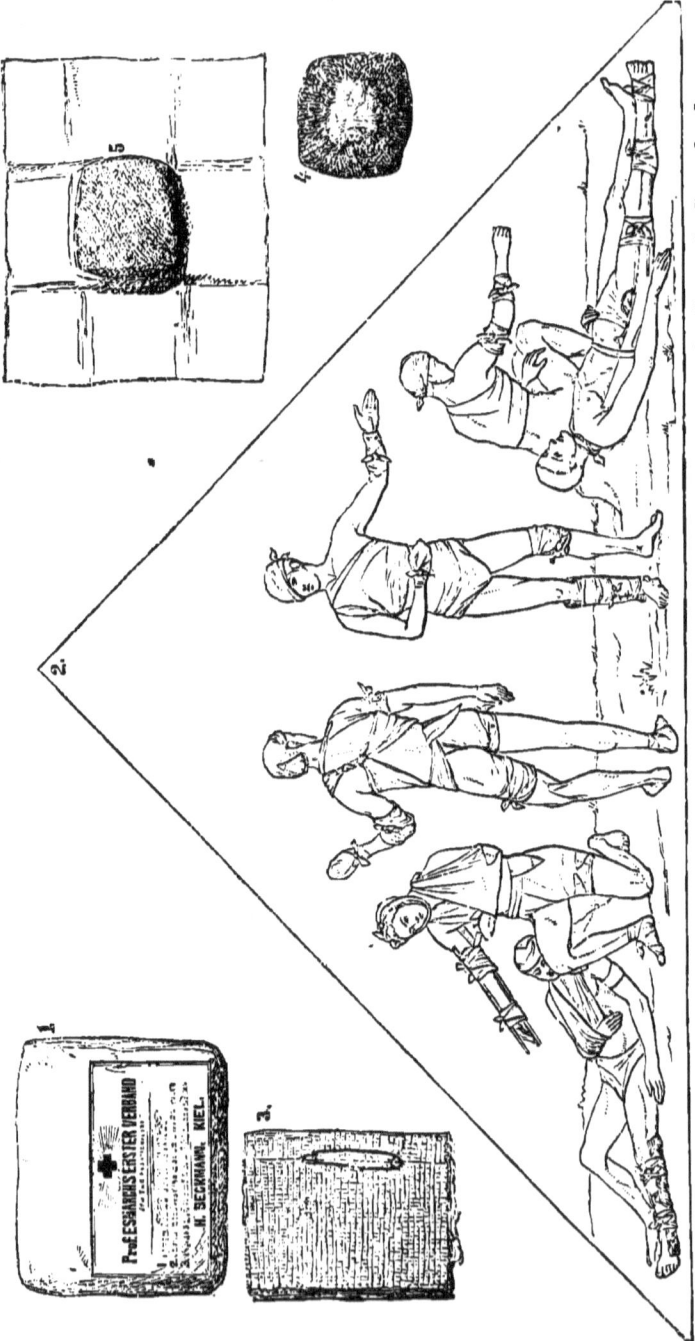

Fig. 60.—Esmarch's first dressing for wounded in battle. 1. Packet folded up. 2. Triangular bandage. 3. Gauze bandage. 4. Antiseptic tampon. 5. Tampon and square of oiled paper.

septic, and may be applied in the same way over the wound, and fastened by a bandage similarly prepared (fig. 59).

It is very doubtful, however, how far soldiers can be induced to carry such dressings about with them, or that they will be clean enough for use when the time comes. The members of the ambulance corps might have a supply ready prepared for application, and be instructed how to use them.

The tampon should be invariably applied directly to the wound, and the oiled paper outside; were this interposed between the tampon and the skin, the chance of a scab being formed would be prevented.

The object of the triangular bandage is chiefly to afford support, or to procure the immobilisation of the injured limb or part.

It would be quite impossible to analyse the large number of contributions made on the employment of antiseptic treatment in military surgery; one, however, I may allude to as showing the success which may be accomplished by secondary antiseptic methods, as I have already referred to Reyher's brilliant results obtained by primary antiseptic treatment.

Dr. Cammerer, Surgeon-General of the German army, has published an account of observations in Roumania during the Russo-Turkish war.[1] The cases came from Plevna, and were treated by Prussian army-surgeons under his direction. They had been subjected to three, four, and five days' previous transport, and arrived at the Roumanian hospitals with wounds swollen and inflamed. Yet, although an interval of seven days from the receipt of the injury had occurred in most of the cases, these septic wounds were converted into aseptic ones by repeated washings out with 5-per-cent. carbolic solution, and by the external application

[1] *Deutsche Militairärztliche Zeitschrift*, July 1878.

of moist carbolic jute dressings. The surface of the part injured was first washed with soap and carbolic, shaved, and again washed. All foreign bodies were removed from the wound, ill-conditioned granulation-tissue scraped away, the 8-per-cent. solution of chloride of zinc applied, followed by careful drainage and jute-dressing externally, which was renewed at first every twenty-four hours. Although large wound-cavities were washed out with the 5-per-cent. carbolic solution, no toxic effects were observed to follow.

When Dr. Cammerer took charge of the hospital he found in it cases of pyæmia, septicæmia, typhus, and tetanus. Afterwards no case of septic poisoning occurred, and the wounds healed well without fever and with but slight discharge.

All the surgeons unite in stating that after an interval of even so long as fourteen days gunshot wounds may be made aseptic.

The interval, indeed, seemed to be of less importance than the configuration of the wound, and the power of thoroughly disinfecting and efficiently draining it. When this could be done the wounds, without exception, became aseptic after the third, fourth, or fifth dressing.

The intermediate condition of diffuse inflammatory purulent œdema, where no distinct abscess-cavity has formed, is that in which asepsis is most difficult to obtain, and probably the great success of Dr. Cammerer's staff is due to the circumstance that most of the cases of this character had been previously eliminated by death. All the wounds of the soft parts, with few exceptions, did well under the treatment described. Three wounds which involved the knee, and two which implicated the elbow-joint, recovered like simple wounds of the soft parts; one of the former cases was a perforation by bullet, *loch schuss*, of the condyles of the femur. In six weeks the patient was able to get about with a movable joint, and only a small superficial surface to heal. In a second instance the bullet remained lodged in the bone, but the patient recovered in a month with a movable knee-joint. The third made a similar recovery after a transverse wound through the superior synovial pouch.

The treatment consisted in making free incisions into the articulation, washing it out with strong carbolic solution and applying moist jute.

Laué, speaking from an extensive experience in the hospital at Königsberg,[1] praises Münnich's dry carbolic jute for use in war on account of its facile perforation, efficiency as an antiseptic, its cheapness, which is about half that of gauze, so that it can be destroyed at once as soon as used; the latter an absolute necessity, I think, for every kind of dressing used in war time. It also serves the purposes of sponges. Laué incidentally remarks that while rubber drainage-tubes may not be forthcoming in time of war, abundance of horse-hair always is, and that, failing the one, horse-hair drains may be employed with advantage.

The jute is laid on the surface in place of the gauze in a quite similar fashion, all the other details being the same as with the gauze dressing. The extent of surface covered by the jute of course depends on the nature of the case; the thickness varies from three to five centimètres. For use in military surgery it may be compressed, and carried in tin cases. Its cheapness; the ease with which it may be freshly prepared day by day, even by the least skilled assistant; the facility with which it can be separated into small or large masses for various purposes, together with its proved antiseptic power, are weighty recommendations. Bruns' gauze possesses similar advantages, and is more convenient as an application to wounds, but the greater cost is an objection to it.

[1] *Deutsche Militairärztliche Zeitschrift*, May 1879.

A detailed account of many interesting cases is also given in Dr. Cammerer's paper. For instance, a soldier who was wounded at Rahova by a bullet which traversed his thigh, and fractured the femur at the junction of the upper and middle third. He was jolted along in an ox waggon for four days till he reached Turnu Magurelli. Yet with secondary antiseptic treatment the wound healed, and the bone was united in eight weeks.

The admirable results obtained by these German surgeons, in spite of the ridicule with which their Roumanian brethren regarded their proceedings, is worthy of the close attention of our military medical officers, and affords an encouraging prospect for a more successful treatment of gunshot injuries. It shows that, however valuable the primary antiseptic treatment of such wounds may be, a secondary antiseptic treatment can also produce excellent results.

The results of the treatment of gunshot injuries of the knee furnish a crucial test of the value of surgical treatment.

Hennen, Larrey, and Guthrie all agree that gunshot wounds of the knee-joint demand amputation, as the result is otherwise invariably fatal. Guthrie states in his book that he cannot recollect a case of recovery after gunshot fracture of the joint-ends of the bones. Longmore tells us that in the Crimea not a single man wounded in the knee-joint recovered without amputation.

Langenbeck in 1868 first propounded the contrary opinion that in gunshot fracture involving the knee-joint the best results were attained by conservative treatment, unless the bone-injury were very extensive. In the wars of 1866 and 1870–71 he put his opinions to the test. The success, however was not very great, and Heintzel's published experience[1] of the campaign of 1870 resembles that of many

[1] *Deutsche Militairärztliche Zeitschrift* July 1875, p. 305.

others. He says that of 44 gunshot fractures of the knee treated conservatively 17 recovered and 27 died, a mortality of 61·3 per cent. Nevertheless it came to be recognised that a wound implicating the capsule alone, or to a small extent the bone, was best treated by the expectant method. There is, of course, in these cases great difficulty in arriving at an exact diagnosis as to the precise extent of the damage.

A most important peculiarity in knee-joint wounds is the probability of their becoming valvular, and to that extent subcutaneous. For instance, the knee-joint while flexed may be traversed by the bullet, and so soon as the limb is straightened the internal and external openings cease to correspond. Experience proves how often severe gunshot injuries of the knee have healed, and often, probably, it may be, under conditions such as these.

Bergmann, formerly consulting surgeon to the Imperial army of the Danube, gives a striking illustration of the fact in an interesting discourse [1] delivered on his installation as professor in Wurtzburg. After the storming of Telisch and Gorni Dubnik 15 cases of compound fracture of the knee-joint came under his treatment, mere injury to the capsule being excluded from the list. Of these 14 recovered, 2 after amputation, and 1 died, also after amputation. In 5 the bullet was impacted, being afterwards removed in 3 instances.

The treatment consisted in the disinfection of the exterior of the limb, the application of a quantity of salicylic wool to the wound, and a plaster-of-Paris splint which included both the hip and ankle joints. No examination of the interior of the wound was ever made until the attempt to procure healing in the manner described had failed.

[1] *Ueber die Behandlung der Schusswunden des Kniegelenks im Kriege.* Stuttgart, 1878.

The question of excision in time of war is not so materially changed by the introduction of antiseptic treatment as are many other operative methods. Incisions into joints, however, will largely take the place of excisions.

Hitherto secondary resections have afforded the best results, and the operation has been performed either to remove seriously injured bones, or to act as an antiphlogistic, and give better drainage, when the inflammation runs high.

But with antiseptic precautions primary resection may probably become a successful operation, although the circumstances under which it is indicated are considerably narrowed. Certainly such operations should never be performed, in the first dressing lines, as I have known them to be done, and done extensively, but only where the means for a sufficiently careful after-treatment are afforded.

Amputation is another operation becoming less and less frequent as improvements in the methods of dressing wounds progress. But many cases, such as the complete destruction of the limb by shell explosion, and injury involving both vein and artery, must unfortunately be treated in this radical way; also when long and difficult transport, or some other circumstance, renders a conservative treatment impossible. A primary amputation should always be performed in case of necessity rather than a secondary one.

Hueter remarks that the results of early, as compared with late amputations are very similar in effect to what happens to a drowning man according as he may be pulled out of the water one minute or ten minutes after his immersion. The earliest moment should be chosen for operation in cases demanding amputation.

METRIC TABLE.

As the metric system has been frequently made use of, the following tables may prove convenient for reference.

Metric Measures of Length.

1 Millimètre	0·001 =	·039 inches
1 Centimètre	0·01 =	·393 ,,
1 Mètre	1· =	39·370 ,,

Metric Weights.

1 Milligramme	0·001 =	$\frac{1}{64}$ grains
1 Centigramme	0·01 =	$\frac{1}{6}$,,
1 Gramme	1· =	15·432 ,,
1 Kilogramme	1000 =	2·7 lbs.

Approximate Equivalents.

1 Minim or 1 Grain	=	·06 grammes
1 Drachm	=	4· ,,
1 Fluid Ounce	=	30· ,,
1 Ounce	=	31· ,,

Equivalent Temperatures.

Normal Temperature	98°·4	Fahr.
37° Cent.	98°·6	,,
38° ,,	100°·3	,,
39° ,,	102°·2	,,
40° ,,	104°·	,,
41° ,,	105°·8	,,

One may easily convert Centigrade into Fahrenheit by rule of three; that is, Centigrade is to Fahrenheit as 5 is to 9, adding 32° to the result.

Reamur or Celsius is to Fahrenheit as 4 is to 9.

INDEX.

ABD

ABDOMINAL section, 203
    for intestinal obstruction, 204
    Risks of, 205
Abscess, 175
    Axillary, 176
    Drainage of, 176
    Fistulæ after, 179
    Lumbar, 176
    Mammary, 176
    Pelvic, 178
    Psoas, 176
    Strumous, 178
    Treatment of, 175
Absorption by granulations, 134
Absolute phenol, 116
Acetate of alumina, 133
    Preparation of, 133
    Uses of, 133
    Maas' results with, 8, 133
Action of spray, 159
Address, 1
Adjuncts to the antiseptic treatment, 100
Air-dust, 104
Amputation, 18
    of thigh, 164
    Results of, 18, 26, 29, 31, 37
    After gun-shot injury, 280
    Schede's comparison of results, 31
Amputation of the breast, 236
    Drainage of wound, 237
    Dressing of, 237
Anchylosis of joints, 196
Anterior curvature of tibiæ, 180
Antiseptics in military surgery, 265
    Difficulties of application, 265
Antiseptics in private practice, 10
Antiseptic practice, 162
    Objects of, 4, 162
Antiseptic theory, 100
Arrest of hæmorrhage, 164
Aseptic wound fever, 101
Atresia vaginæ, 239
    Operation for, 239

CAR

BANDAGES, 137
    Elastic, 138
    Gauze, 137
    Bardenheuer's results, 29
    Boracic acid, 131
    Lotion, 131
    Lint, 132
    Ointment, 132
    Vaseline, 132

CANCER of the rectum, 246
    Excision of, 247
    Indications for, 246
    Methods of operating, 247-249
    Leaving anus, 247
    Dilating sphincter, 247
    Dividing sphincter, 247
    Drainage of wound, 248
    Dressing of wound, 250
    Wound of peritoneum, 249
    Permanent irrigation, 250
    Stricture after, 252
    Results of, 251
Carbolic acid, 106
    Action of, 102, 116
    Bruns' concentrated mixture, 123
    Ill-effects of carbolic acid, 111
    Impurities of, 116
    Preparation of, 116
    Solutions of
        Alcoholic, 119
        Oily, 118
        Watery, 117
    Carbolic vaseline, 10
Carbolic acid poisoning, 124
    Symptoms, 125
    Test for in Urine, 126
    Treatment, 126
Carbolic eczema, 126
Carbolic erythema, 126
Cartilage, changes in, after a disarticulation, 173
Cathetcrisation, 241
Cleansing of catheters, 242

Chloride of zinc, 134
  Uses of, 134
  Jute, 135
Clamp artery forceps, 165
Comparative statistics in Glasgow Infirmary, 98
Compound fracture, 103, 180
  Causes of death after, 180
  Treatment of, 180
  Counter-openings, 181
  Delayed and non-union, 183
  Disinfection of wound, 180
  Drainage of, 181
  Dressing of, 183
  Mode of healing of soft parts, 183
  Plaster splints for, 183
  Removal of bone, 181
  Results in, 13, 16, 17, 26, 27, 97
Course of an antiseptically dressed case, 172
Cresol, 116
Cystitis after catheterisation, 242

DEBATE, 47
  Mr. Morrant Baker, 94
  Mr. Barwell, 53
  Mr. Bryant, 47
  Mr T. Holmes, 57
  Mr. J. Hutchinson, 77
  Prof. Lister, 59
  Mr. Lund, 88
  Mr. Mac Cormac, 96
  Mr. Macnamara, 49
  Dr. Newman, 89
  Sir James Paget, 81
  Mr. T. Smith, 56
  Mr. J. K. Thornton, 91
  Mr. Spencer Wells, 54
  Prof. J. Wood, 71
Decalcified bone tubes, 149
  Absorption of, 149
  Dressing used with, 150
  Preparation of, 149
  Results with at Kiel, 23
Diffuse phlegmon, 263
  Treatment of, 263
  Irrigation, 266
  Suspension of limb, 265
Disuse of spray, 160
Drains
  Drainage, importance of, 5, 146
  Horsehair, 150
  Catgut, 151
Drainage-tubes
  India-rubber, 147

Drainage-tubes, *continued*
  Uses of, 146, 147
  Preservation of, 148
  Securing in wound of, 148
  Time for removal of, 147
  Double-barrelled, 176
  Decalcified bone, 149
Dressings
  Indications for changing, 171
  Elements of failure, 103, 120
  Method of performing, 169
  Necessaries for, 173

EMPYEMA, 234
  Case, 236
  Drainage of pleura, 235
  Excision of rib, 236
  Mode of opening, 234
Enchondroma of chest-wall, 253
Esmarch's first dressing for the wounded in battle, 271
Excision of pylorus, 205

FÆCAL fistula, 206
  Cases, 206
  Dupuytren's clamp, 206
  New operation for, 207
Faulty union after fracture, 196
Filo-pressure, 141
Fistulæ, treatment of, 180
Fractured patella, 193
  After-treatment, 195
  Mode of union, 193
  Methods of treatment
    By Malgaigne's hooks, 197
    By paracentesis of joint, 195
    By suture, 194
    Result of case, 195

GASTROTOMY, 204
  Gauze
    Carbolised, 119, 122
    Bruns', 6, 122, 126
    Küster's, 121
    Lister's, 119
    Retention of carbolic acid by, 120, 124
    Thymol, 131
  Genu valgum, 189
    Pathology of, 191
    Result of osteotomy for, 11, 193
    Treatment of, 190
  Germ theory, 104
  Gun-shot injuries, 266
    Amputation for, 280
    Antiseptic tampon for, 269–272

INDEX. 285

Gunshot-injuries, *continued*
  Drainage of wounds, 271
  Dressing, 271, 273
  Examination of wounds, 268
  Purulent œdema after, 276
  Resection for, 280
  Results of, 38, 41, 274, 278
  Reyher's statistics, 41–44
  Septic disease after, 43
  Subcutaneous nature of, 268

HEAD injuries, 219
  Cases, 222
  Mode of dressing, 221
  Results, 40
  Tolerance of brain-substance, 219
  Trephining, 219, 220
Healing under antiseptic dressings, 113
Healing beneath a scab, 112
Hernia, radical cure of, 30, 209
  Aim of operation, 211
  Cases, 208
  Congenital, 210
  Femoral, 210
  Inguinal, 209
  Results of operation, 30, 211
Hot water as a hæmostatic, 167
Hydatid of liver
  Operation for radical cure, 218
  Results of, 219
Hydrocele, radical cure of, 243
  Advantages of antiseptic incision, 245
  Mode of dressing, 244
  Modified operation, 246
  Retention of urine after, 244
  Volkmann's operation for Hydrocele, 243

ILLUSTRATIVE amputation, 164
  Influence of antiseptic treatment on suppuration, 110
Influence of the atmosphere on wounds, 104

JOINTS, puncture of, 201
  Compound fracture into, 28
Jute, 127
  Carbolised, moist, 124, 125, 127
    dry, 127
  Chloride of zinc jute, 135
  Neuber's jute dressing, 150
  Retention of carbolic acid by jute, 127
  Salicylic jute, 130

KNEE-JOINT
  Drainage of, 18, 203
  Gun-shot wound of, 39, 44, 271, 278
  Treatment, 277
  Results after injury, 39, 41, 44, 278
  Loose cartilages in, 19
  Results of direct incision, 19, 20

LIGATURE of blood-vessels, 232
  Advantages of catgut, 233
  Lister's experiments, 232
Ligatures
  Catgut, 138
  Absorption of, 140
  Advantages of, 140
  Preparation of, 139
  Preservation of, 138
  Silk, 142
  Preparation of, 142

MACKINTOSH, 136
  Substitutes for, 137
  Uses of, 136
Metric table, 281
Micro-organisms in air, 105
  Beneath antiseptic dressings, 108
  In wounds not treated antiseptically, 109.

NATURE of blood-poisoning, 2
  Neuber's drains, 7
Nerves, suture of, 255
  Transplantation of, 256

ORGANISATION of blood-clot, 115, 272
Osteotomy, 188
  Results, 10, 27, 50, 189
Ovariotomy, 212
  Cleansing of peritoneum, 215
  Dressing, 215
  Ligature of pedicle, 214
  Mode of operating, 214
  Mortality after, 212
  Septic peritonitis after, 213
  Results, 20, 21, 22, 212

PARAFFINE, 120
  Parenchymatous hæmorrhage, 167
Plaster-of-Paris splints, 183
Preparations for an operation, 162
  Necessary items, 164

PRE

Prevention of wound-diseases, 102
Process of repair in wounds treated antiseptically, 113
Protective, 135
   Preparation, 135
   Discoloration of, 136
Puncture of joints, 201
   After-progress, 203
   Cases suited for, 202
   Mode of operating, 202
   Results, 18, 203
Putrefaction and wound-diseases, 3, 105

RELATION of blood-poisoning to wounds, 3
Resection, substitution of use of sharp spoon for, 180, 203
Resection on battle-field, 280
Resection of shoulder, 238
Resin, 120
Resistance of healthy living tissue to germs, 107
Roughness of hands due to use of carbolic acid, 8

SALICYLIC acid, 128
   Solutions of, 128
   Detection of in urine, 130
   Estimation of strength of wool, 129
   Jute, 130
   Preparation of, 128
   Wool, 129
   Uses of, 129
Septic wounds, 260
   Disinfection of, 262
   Management of, 260
   Use of sharp spoon, 261
Skin-grafting, 132
Sponges
   Disinfection of, 151
   For ovariotomy, 152
   Ill-effects of hot water on, 152
   Preparation of, 151
   Preservation of, 151
Spray producers, 153
   Foot, 154
   Hand, 153
   Richardson's, 154
   Steam, 155

WRI

Statistical results, 9
Strumous joints, 203
   Treatment of, 203
Sutures
   Catgut, 141
   Horsehair, 142
   Insertion of, 143
   Leaden plate, 144
   Silk, 142
   Silkworm gut, 142
   Thiersch's bead, 145
   Wills' button, 145
   Wire and button, 144
   Wire, 142
Suture
   Of Nerves, 255
   Musculo-spiral, 255
   Of tendons, 257
   Cases, 259
   Mode of operating, 257
   Tendo Achillis, 259
   Wrist, 258

TENDON
   Suppuration in sheath of, 264
Thymol, 131
   Inefficiency of, 8, 131

UNUNITED fracture, 197
   Deficient local reaction, 197
   Lister's results, 197
   Of femur, 199
   Of neck of femur, 200
   Of lower end of femur, 199
   Of patella, 194
   Use of pegs, 198
   Use of iron nails, 199

VAGINAL operations
   Method of dressing, 241
Varicose veins,
   Excision of, 234
Vicious union of fracture, 197

WOUND, diseases
   At Halle, 26
   At Munich, 22
   Exclusion of, 2, 22, 162
   Frequency of, 102
Wrist-joint, wound of, 258

SMITH, ELDER, & CO.'S PUBLICATIONS.

SURGERY: its Principles and Practice. By TIMOTHY HOLMES, M.A. Cantab., F.R.C.S., Surgeon to St. George's Hospital. Second Edition. With upwards of 400 Illustrations. Royal 8vo. 30s.

ELEMENTS of HUMAN PHYSIOLOGY. By Dr. L. HERMANN, Professor of Physiology in the University of Zurich. Second Edition. Entirely recast from the Sixth German Edition, with very copious additions, and many additional Woodcuts, by ARTHUR GAMGEE, M.D., F.R.S., Brackenbury Professor of Physiology in Owen's College, Manchester, and Examiner in Physiology in the University of Edinburgh. Demy 8vo. 16s.

A SYSTEM of SURGERY: PATHOLOGICAL, DIAGNOSTIC, THERAPEUTIC, and OPERATIVE. By SAMUEL D. GROSS, M.D., LL.D., D.C.L. Oxon. Fifth Edition, greatly Enlarged and thoroughly Revised, with upwards of 1,400 Illustrations. 2 vols. 8vo. £3. 10s.

The ESSENTIALS of BANDAGING: for Managing Fractures and Dislocations; for administering Ether and Chloroform; and for using other Surgical Apparatus; and containing a Chapter on Surgical Landmarks. By BERKELEY HILL, M.B. Lond., F.R.C.S., Professor of Clinical Surgery in University College, Surgeon to University College Hospital, and Surgeon to the Lock Hospital. With 134 Illustrations. Fourth Edition, Revised and much Enlarged. Fcp. 8vo. 5s.

SPINAL DISEASE and SPINAL CURVATURE: their Treatment by Suspension and the Use of Plaster-of-Paris Bandage. By LEWIS A. SAYRE, M.D., of New York, Professor of Orthopædic Surgery in Bellevue Hospital Medical College, New York, &c. &c. With 21 Photographs and numerous Woodcuts. Crown 8vo. 10s. 6d.

A HANDBOOK of OPHTHALMIC SURGERY. By BENJAMIN THOMPSON LOWNE, F.R.C.S., Ophthalmic Surgeon to the Great Northern Hospital. Crown 8vo. 6s.

The STUDENT'S MANUAL of VENEREAL DISEASES. Being a concise description of those Affections and of their Treatment. By BERKELEY HILL, M.B., Professor of Clinical Surgery in University College, London; Surgeon to University College and to the Lock Hospitals; and by ARTHUR COOPER, late House Surgeon to the Lock Hospital. Second Edition. Post 8vo. 2s. 6d.

SKIN DISEASES: including their Definitions, Symptoms, Diagnosis, Prognosis, Morbid Anatomy, and Treatment. A Manual for Students and Practitioners. By MALCOLM MORRIS, Joint Lecturer on Dermatology, St Mary's Hospital Medical School; formerly Clinical Assistant, Hospital for Diseases of the Skin, Blackfriars. With Illustrations. Crown 8vo. 7s. 6d.

ESSENTIALS of the PRINCIPLES and PRACTICE of MEDICINE. A Handbook for Students and Practitioners. By HENRY HARTSHORNE, A.M., M.D. New Edition. 12s. 6d.

A TREATISE on the THEORY and PRACTICE of MEDICINE. By JOHN SYER BRISTOWE, M.D. Lond., F.R.C.P., Physician to St. Thomas's Hospital, Joint Lecturer in Medicine to the Royal College of Surgeons, formerly Examiner in Medicine to University of London, and Lecturer on General Pathology and on Physiology at St. Thomas's Hospital. Second Edition. 8vo. 21s.

London: SMITH, ELDER, & CO., 15 Waterloo Place.

# SMITH, ELDER, & CO.'S PUBLICATIONS.

DEMONSTRATIONS of ANATOMY; being a Guide to the Knowledge of the Human Body by Dissection. By GEORGE VINER ELLIS, Professor of Anatomy in University College, London. Eighth Edition. Revised. With 248 Engravings on Wood. Small 8vo. 12s. 6d. The number of illustrations has been largely added to in this edition, and many of the new woodcuts are reduced copies of the Plates in the Author's work, 'Illustrations of Dissections.'

ILLUSTRATIONS of DISSECTIONS. In a Series of Original Coloured Plates, the Size of Life, representing the Dissection of the Human Body. By G. V. ELLIS and G. H. FORD. Imperial folio, 2 vols. half-bound in morocco, £6. 6s. May also be had in parts, separately. Parts 1 to 28, 3s. 6d. each; Part 29, 5s.

MANUAL of PRACTICAL ANATOMY. With Outline Plates. By J. COSSAR EWART, M.D. Edin., F.R.C.S.E., F.R.S.E., Lecturer on Anatomy, School of Medicine, Edinburgh. Part I. The Upper Limb. Demy 8vo. 4s. 6d.

The EXAMINER in ANATOMY: A Course of Instruction on the Method of Answering Anatomical Questions. By ARTHUR TREHERN NORTON, F.R.C.S., Assistant-Surgeon, Surgeon in Charge of the Throat Department, Lecturer on Surgery, and late Lecturer on Anatomy at St. Mary's Hospital, &c. Crown 8vo. 5s.

A DIRECTORY for the DISSECTION of the HUMAN BODY. By JOHN CLELAND, M.D., F.R.S., Professor of Anatomy and Physiology in Queen's College, Galway. Fcp. 8vo. 3s. 6d.

MANUAL of PRACTICAL and APPLIED ANATOMY, including Human Morphology. By H. A. REEVES, F.R.C.S., late Demonstrator of Anatomy at the London and Middlesex Hospitals, Assistant Surgeon and Teacher of Practical Surgery at the London Hospital, &c. Vol. I. With numerous Illustrations. 8vo. [In the press.

QUAIN and WILSON'S ANATOMICAL PLATES. 201 Plates. 2 vols. Royal folio, half-bound in morocco, or Five Parts bound in cloth. Price coloured, £10. 10s.; plain, £6. 6s.

A COURSE of PRACTICAL HISTOLOGY. By EDWARD ALBERT SCHÄFER, Assistant Professor of Physiology, University College. With numerous Illustrations. Crown 8vo. 10s. 6d.

246 Outline Drawings with adhesive backs, for Clinical Case Books.
OUTLINE DIAGRAM FORMS for CLINICAL CASE BOOKS. For the representation of Injuries and Diseases and Physical Signs. Designed for the use of Clinical Students, Physicians, and Surgeons. By G. ROWELL, M.D., Resident Surgeon to the Leeds Infirmary. 3s. 6d.

The NOTATION CASE BOOK. Designed by HENRY VEALE, M.D., Assistant Professor of Military Medicine in the Army Medical School; Surgeon-Major, Army Medical Department, &c. Oblong crown 8vo. for the pocket, 5s.

London: SMITH, ELDER, & CO., 15 Waterloo Place.

www.ingramcontent.com/pod-product-compliance
Lightning Source LLC
Chambersburg PA
CBHW032045230426
43672CB00009B/1479